Changing Pattern of Land Use in Jharkhand

Dr. Shyam Nandan Sinha

RAJESH PUBLICATIONS
NEW DELHI - 110 002

ISBN : 978-81-94426-95-0

Published by :
Sanjay Gupta
RAJESH PUBLICATIONS
1, Ansari Road, Darya Ganj
New Delhi - 110002
Ph.: 011-23274550

Printed in India by :
Belwel Process
Delhi-110002

First Published in India 2020

Price : Rs. 800.00

Foreword

This book entitled "Changing Pattern of Land Use in Jharkhand" has grown out the Ph.D. thesis submitted in 2019 to the Vinoba Bhave University, Hazaribag (Jharkhand) by Dr. Shyam Nandan Sinha. This is undoubtedly a detailed and meticulous study on changing pattern of land use in the study area.

This book is pioneer work by the author who has describe the factors for land use in the area. This book is informative, stimulating and productive and deserves to be read by a wide audience. I hope book will be find favour with all those who are interested in the study of the land use planning.

Dr. Kamla Prasad
Former Head, Department of Geography
Vinoba Bhave University,
Hazaribag-825301 (Jharkhand)

Preface

The present work is the outcome of an intensive field work of land use of Dhanbad district of Jharkhand. It deals with different aspects of land use of an area which has known as coal capital of India. The area lies in north portion of Chhotanagpur plateau. It covers an area of 2040 sq.km where a little above 26 lakh people live. The area presenting a distinctive geographical personality. It has supported a dence population. Our landmass is fixed, but the competition among different kinds of uses for this land is increasing because of rapidly growth of population. Therefore, integrated land resource planning and management are required to resolve these conflicts and soil resource survey seems to be a viable means in this process and knowledge of soil fertility status and problem of soils become essential for sustainable land use plan.

One cannot think of progress without a proper utilization of land. Hence the study of land use of Dhanbad district is of immense value in any scheme of regional development and planning. The present work is an attempt to portray and analyse the land use pattern of an important areas based on first hand study of eight sample villages readily selected from each of the eight development blocks that together constituents the district of Dhanbad in Jharkhand.

The present work has been divided into six chapters. The first chapter explains the concept of land use, and its significance. It further deals with the objective of land use studies, review of literature and purpose of study the research design, sources of data, methodology and sampling techniques. The geographical personality is presented in the second chapter. Here the physical and socio economics background have been analysed. Chapter third deals with changing pattern of general land use. The case study of eight villages are described in the fourth chapter. The last two chapters i.e. fifth and six deals with analytical frame work and conclusion respectively. A brief summary is also presented at the end.

Illustration have been selected for their direct and clear impact. Photographs have been carefully selected to illustrate precisely the subject under discussion. At the each chapter reference have been mentioned and a bibliography has been placed at the end of the discussion.

Dr. Shyam Nandan Sinha

Acknowledgement

In the completion of this work I had to take help from my Teachers, colleagues, friends and government officials etc. I must express my sincere thanks to all of them.

First of all I offer my sincerest and deepest sense of gratitude to my supervisor Dr. O.P Mahto, university professor and ex. head of the post graduate department of Geography, Vinoba Bhave University, Hazaribag, who was kind enough to supervise the entire work and give valuable suggestion.

I would like to offer my thanks to my teachers and ex. head, Dept of Geography, V.B.U. Hazaribag Dr. Kamla Prasad for their help at different stages.

I am also thankful to my friends Dr. A.K. Ramanuj, Asst. Professor of Geography, Barhi College, Barhi and Dr. Tikait Manjhi, Asst. Professor of Geography, K.S.G.M Nirsa and villagers who gave company in the field.

I would like to offer my deep gratitude to Dr. J. Bhagat, Dr. Pradeep. K. Singh, Dr. K.K. Singh, Dr. S.R. Sharma, Prof. A.K. Sinha and many others of V.B.U Hazaribag for inspiration to complete this task.

I would like to express my thanks to my wife Smt. Arpana Sinha, who encouraged me in this work.

Finally heartly thanks are also due to Sh. Sanjay Gupta, Prop. Rajesh Publications, New Delhi for the help and encouragement we received to get their work published.

Dr. Shyam Nandan Sinha

Keywords

AGRICULTURE

BAHAL

DENSITY

DHARWAR

IMPORT

IRRIGATION

KANALI

OCCUPATIONAL

PASTURE

UTILIZATION

Contents

List of Figures

List of Tables

List of Illustrations

1
Chapter **Introduction**

Land use is the surface utilization of all kinds of lands, at a given time and space. This leads one back to the village farm and the farmer to the field, gardens, pastures, fallow land, and forests and to the isolated farmstead. The role of geographers becomes to analyze[1] the relationship between various uses of land and planning. This analysis enables people to use the land more properly and obtain high yield to solve the problem of man.

The primary uses of land are for crops, forest, pasture, mining, transportation, garden, residential, recreational, industrial, commercial and uncultivable waste, barren and fallow land. "Land use is also related to conversion of land from one major use to another[2] general use". The use of land changes according to the changing needs of man. Stamp has classified these need of man into six categories. The need of work, home, food, transportation, communication, defense and[3] recreation.

To meet needs the man has to look towards land .The need of food means conservation of good agricultural land for production and at the same time the development of poorer land for production of food and cash cops. The same land has to satisfy the need the need of transportation and communication by means of railway, highways and airways. Land is also given to recreational usage in shape of parks, playground, garden, clubs. Apart from these, a large number of uses may be listed.

Land is the expression of a variety of topographical features, natural drainage and basins together with evergreen, deciduous, and mixed forests and soft carpet of tall, small, wet, and dry grasses. Man on the other hand ingraves high agricultural farms and fields, constructs his home, ranging from hamlet to sky high building and ultimately nets his rural and urban settlements with lines of communication. Cultural landscape thus emerges when he depletes forests, till cleared lands and taps underground resources. This relationship between man and land portrays a variety of land problems which differs over time and space.

In this effort to use the land, man has made proper uses misuses, under uses and over uses. When the land is given to a particular use the interest of other uses may be overlooked. The factories are located without any regards to the agricultural value of land. In the same way, the growth of cities in most cases takes place in rich agricultural land. In China, the best agricultural land is converted into graveyard as people try to bury the important person in the best agricultural field to honour him. The tendency leads to grave danger. Land that has been once drawn from agricultural use seldom returns to[4] the former, and if it does, it will only be after a long lapse of time.

Again there are conflicts within agricultural itself and between the use of land for cash crops and food crops. The growth of population and the resultant needs make the conflict of land use more serious. The increasing population should be followed by rise in food production. But as the land is limited, it has to be used very carefully and judiciously. If the food growing field are intensively cultivated without proper manuring, the food obtained from such field would be deficient in mineral, fats, proteins. Which are very useful for our physical and mental well being.

The concept of land use capacity refers to the ability of any given unit of land resource to produce a net return above the production cost associated with its use. The amount of this net return provides an index of use capacity. Land areas with high uses capacity normally have higher market values than those of lower used capacities. Farmer tends to use their land resources for those purpose. Which promise them the highest return. In fact a land use study should find measures to provide optimum return to land utilizes. A land utilizations project aims and striking balance between added mouths and land use capacity. The concept of land use, therefore, revolves round the man's accomplishments in conversion of land, major use to another general use. Each stage of such change may involve many problems to[5] pave the path for attaining equilibrium in use of land.

Increasing population and changing needs of the time requires revision of land utilization the revision is done by trial and error method, which leaves its trace of success and failure. The study of land utilization makes it imperative to present an excellent opportunity to rectify past error and to overrules further errors through scientific methods. The success of national planning is dependent upon the proper utilization of land. Some day in our country a planned programmed will determine the pattern of land use and there not only crops and tamed animals but indirectly things will be determined by man's conscious[6] planning and use of land.

The demand of land changes due to changing needs of society. As socio-economic conditions change, land use keeps on changing. The study of land use, therefore, a subject of continuous interest.

REVIEW OF LITERATURE :

The study of land use is important not only in agriculture dominated region but throughout the world because of its relationship with different human phenomena. Its significance has multiplied during recent year due to scientific innovations and increased pressure on land. During early year there was enough land to support to limited population. But today population explosion has remarkably reduced the man-land ratio. Because of increasing pressure on land, intensive and proper used land have become essential. Geographers, planners, and other scholars have realized the importance of the subject and they have started paying more attention to the problem and other scholars have realized the importance of the subject and they have started paying more attention to the problem.

Stamp pioneered the land use study in Britain. In the year 1930, he established an independent research organization called "land utilization survey of Britain". The main motive of the formation of the organization was to prepare land use maps of Britain. For this purpose, intensive fieldwork was conducted and different land use map were prepared. For fieldwork, map of the size of 18" x 12" were made available because such map were quite handy. A large number of students from schools, colleges and universities were entrusted for this work. Ultimately he could prepare the land use map before the out break of first world war. On the basis of these maps a voluminous book entitled 'The land of Britain; its use and misuse" was published in the year 1962'.[7]

The land use work of stamp become the guideline for researchers not only in Britain but all over the world. Today geographers of developed and developing nation has been trying to make land studies based on this guidelines. Certainly very few nations have been able to prepare land use map.

In the United States also several land use studies have been made. The credit of such study goes to Baker who published an article entitled, 'Land Utilization in the United States: Geographical aspect of the problem'. In this paper he has depicted the trends in land utilization and emphasized the need of land classification and survey. Besides this paper government also encouraged the land use studies in the country. Although the program of land uses survey was launched in the year 1935, it was properly executed only after 1938. Nearly one thousand communities were organized and 1,40000 individuals mostly[8] layman, took part in the work. This organization under took the studies of land use including map making and their analyses. Special attention was laid on kind of farming, size of farms in each land use area, area to be recommended for forestry, wildlife, recreation, settlement etc. The problems thus studies suggested plans in every branch, for example farming and farm sizes, land recovery and settlements.

Thus, American land use survey was taken up mainly for the purpose of land use planning on the country- wide scale that embraced the planning of both the physical and human phenomena. But this much is sure that what is proposed in land use planning is as broad as country life and organization should, therefore begin with most general functional organization that can be discovered in an area. Land use studies in U.S.A., on the other hand, were made with the purpose not only to select certain areas for proper use but also to train the students the method of research in land utilizations, to make available the knowledge of Chinese agriculture, and to arouse interest among people of other countries in China's welfare, land utilization, food and population problems. However, this can be remarked that U.S.A has not made so remarkable progress in the field of land use survey like Britain.

While coming to land use studies in other countries it should be noted that very few countries is should be noted that very few countries hold significant position in this respect. Certainly Japan has more enthusiasm in this field. The main motivation factor was how to support a huge population with less cultivated land as such it is not[9] surprising that land survey has been given the highest priority.

Land use surveys are in progress in various countries of the world, particularly in highly crowed countries, where land used planning on a scientific time is vital, i.e. Pakistan, Cyprus, Bangladesh, China, and Poland etc. Land utilization maps Cyprus have been prepared under the supervision of R.R. Rawson and K.R. Sealy, in the department of Geography, London school of economics. The maps of Italy are being published by Italian National Research Society. In Poland, under the direction of J. Kostrowiekai, Department of geography, polish academy of science, developed a new pattern of land utilization based on agricultural typology, agricultural rationalization and programmed[10] agricultural development.

Land use survey was also conducted in China. Data were collected on sampling basis from 22 provinces and on the basis of sample survey data several aspects of land utilization were analyzed such as farms, food, standard of living of the people, marketing and price level etc. However, in the whole study no attempt was made to record the use of land on maps, which is an important aspect in land use study.

As regards land use studies in India it should be noted that several geographers have paid attention on different aspects of land use studies. Certainly most of them have followed the guideline of stamp, in fact Indian geographers got inspiration from him when he attended the 25th session of the Indian science congress held in Calcutta in the year 1938. First of all land utilization survey of 24 Paganas[11]. And Howrah[12] district was conducted by S.P. Chatterjee and[13] thus, in eastern Uttar Pradesh by M.Safi. They have emphasized that land use survey should be carried out combined with the survey of land capability. It will help in determining the best use of land. Chatterjee tried to organize the land use survey in India in 1940 when Geography and Geology sections of Indian science congress Association, he pointed out the necessity of understanding land use survey. The Government of India established a national committee for the purpose under the guidance of Chatterjee. He surveyed 800 villages of West Bengal and brought out 11 land sheets on the scale of 4"=1 mile. The national Atlas organization also, prepared land use maps on a scale of 1:1000000.[14] M. Safi in his paper entitled 'Technique of rural land using planning with reference to India', was of the opinion that land use survey of a vast country like India is easier to be conducted on the basis of sampling as it is very difficult to procure data for all the villages to be surveyed. Moreover, there is a great similarity in the land use pattern of a particular region comprising of several village, so a sample village would do the needful. Shafi has brought out a scheme based on sampling techniques for land use survey of India. He preferred purposive sampling to the other four type, i.e. random, stratified, cluster and systematic. Describing the trend in the study of land use M. Shafi is of the view that the studies are recently shifting towards the application of quantitative technique in the analysis of various land use[15] components. M. Shafi has also assessed the measurement of land resource in terms of food production efficiency per unit area and its conversion in to calories because in his words 'it may not be fair to equate a pound of wheat with a pound of millet as their caloric yield are vastly[16] different' In this way he has studied land capability classification and measures the potential of land after considering the effect of positive and negative variables.[17] E. Ahmad has analyzed land use types in relation physical elements. According to him, slope of the land should be considered in preparing the development scheme of an Indian village. Three consecutive community development blocks i.e Biharsharif, Noorsarai, and Rahui in Nalanda district have been surveyed under the guidance[18] of P. Dayal and A. Sharan. R.N.P. Singh intensively studied the land use of canal-irrigated[19] area of Patna District. K. Z. Amani brought paper on land utilization in[20] village Golgarhi.

Mishra in his studies of land use in khaddar and ravine of the lower middle Gomatri valley has attempted Land use planning for better adjustment of agriculture to the physical environment for[21] optimum exploitation and conservation of natural resource. Ayyar has given statistical approached to the study of crop. Combination regions by suggesting maximum distance method to group important crops[22] into association. Sharma has tested Bennett's method of land capability classification.[23] Lear month and Bhat were of the opinion that sharp regional[24] contrast in agricultural land use needs a spatial analysis. Singh calculated the land capability with potential productive unit and the[25] standard nutrition unit. Das studied the land use of kosi Basin in north[26] Bihar While Roy selected the district of Bhojpur and Rohtas for such[27] study. Rao has emphasized the technique of soil survey for the[28] analysis of land use in the Godavari region Lahari studied the major land use types and agricultural problems for your typical village near[29] Jasidiah. Karim made study of land in Dinapur Ghushapara near Patna.[30] Bhardwaj dealt with land use in the low land of the Beas river in the Bist Jalandhar Doab with particular reference to two villages. Mohalla[31] and Akitunda. Rai made a sample study in Land utilization of five villages in Ballia district and has also contributed a paper on measurement of rural land use in Ajamgarh; middle Ganga Valley Weaver's method of crop combination regions has been applied by Mandal[32], Sinha[33], Tripathi[34] and Garg[35]. Sinha has also taken the help of Nelson's technique while Tripathi and Garg have applied Dol's method[36] also. Jha studied the problem of land utilization in the kosi basin.[37] Yadav has studied the board regional variation in agricultural land use in Rajasthan. Bhatt has suggested the concept of planning from micro village, block and district levels. Indupriya[38], Prasad[39] and Singh have studied the land use problem of different community development block[40] of Bihar. Ataullah, in his studies of urban land use major town along the Ganga in Bihar has differentiated between the urban and rural land[41] uses. Mandal has discussed natural principal, technique models and the role of remote sensing and air photograph in land use analysis.

Different Government agencies have also contributed to the[42] knowledge of land use studies. The "Agricultural Atlas of India," "the National Atlas"[43] and the "Census Atlas"[44] of different states contain choropleth and dot maps relating to land use and crops. Since the ultimate goal of land use study is to suggest the planning for better utilization of available land society and land use planning is not in practice in most areas, scholars should should influence the implementing agencies for its practices, so that proper use of land improved technique of cultivation would not only solve the food problems of the nation, but it would may provide surplus products for export purpose.

PURPOSE OF STUDY :

The aim of the present work is to bring out the salient changes in the land use pattern of Dhanbad district. During post independence years the introduction of scientific method of cultivation, development of irrigation facilities, electrification of the village, increased pressure of population on the land, impact of industrialization and urbanization, have caused considerable changes in land use. The role of land use studies in this way highlight the land use changes which have taken place due to socio economics transformation and administrative set up to suggest measure to increase production by proper uses of land and to use the improved agricultural methods, crops etc. Land use studies also reveal the potentialities of the land, specially soil and suggest land utilization in proper way. Considering all those problem, the present study of Dhanbad district has been selected because it has problem of food shortage and rural poverty, although big part of land is suitable, soil is productive and man power is adequately available.

HYPOTHESIS :

Jharkhand state has been an area of low agriculture productivity. Misuse and under use of land resource is a common feature. As a result of the combined impact of these two poverty and hunger is looming among a very large section of its population. Moreover, the impact of grow more food campaigns IRDP, Green Revolution and other schemes of package programs of agricultural development are very restricted. Large number of cultivation due to many physical and socio-economic factors has not adopted new agricultural systems and hence low agriculture productivity is continuing. In addition to these problems Dhanbad district has experienced rapid urbanization and excessive conversion of agricultural lands into non agricultural uses. Hence the question of maximizing agricultural productivity of limited agricultural land and development of multicropping system is required for sustainable agricultural productivity.

Therefore the hypothesis of the proposed research is as follows :

1. Whether physical and cultural factor have restricted or limited the availability of agricultural lands as per the need of the district?

2. Whether the increasing use of land for non-agricultural purposes is lowering the agricultural productivity?

3. What is the status of irrigation development, is there potential of increasing irrigation for the adoption of multi cropping?

4. What is the condition of other agricultural infrastructure, whether there is any need of improving?

5. Whether misuse and under use of land continues and what has been the trends toward expansion of agriculture?

6. What are the lacunas of cropping pattern, what has been its trends and what should be the best cropping pattern for the area?

7. What is the status of water and soil resource whether these contain potential for developing agricultural land use?

8. And finally what should be the planning for land use development?

SIGNIFICANCE OF STUDY :

The proposed study related to the analysis of the pattern of agricultural land use and assessment of the problems and potentials of developing land use towards optimizing agricultural productivity. In the present state of low agricultural productivity and food crisis of the state, the study of Dhanbad district is very significant. The study will provide plans of optimizing agricultural land use toward the achievement of sustainable agriculture and hence it is relevant for cultivators, government and agricultural planners. Through optimizing agricultural productivity the cultivators may be able to improve their standard of living. The proposed research topic shall provide detail account of agricultural land use of the district where non-agricultural pursuits like mining etc. are more important.

OBJECTIVES OF THE STUDY :

The study of land use has recently developed and it has become one of the most important branches of economic Geography because even today majority of the people have close relationship with the primary activities which concern with land. Geographers can present a clear picture of the potential of land use, conductive to fruitful planning for massive agricultural turn over. A plan for the almost utilization of land asset should essentially include the requirements of land for location of industries, site for houses, school public building, road and railways, irrigational dam and channels, etc. Rural as well as urban areas have land under different uses. In rural areas much of the land is used for agricultural purpose and other uses have less land. On the other hand, in urban areas much of the land remains under residential, commercial, industrial and other uses. Since one use of land exclude another, it is necessary to asses fully the potentialities of every type of land in respect of all possible uses.

The study of land use not only concerns with the land use classification, use and misuse of land, capability, land use planning, but it also deals with several socio economics aspect like man land ratio. Changing pressure of man on land, land use change due to dynamism of socio political condition and scientific innovations etc. As such the study of land use has developed relationship with the study of rural settlements, populations, agriculture science, geomorphology, as well as urban geography. In fact rural and urban land use have become to clear cut aspects of land use study. The study of the settlement either rural or urban would not be perfect without having background information about the land use. Similarly, land use study will require basic information about the land surface and soil characteristics. Demographers do not sit ideal because they establish relationships with land.They analyse pressure of man on land and other aspects. Day by day this aspect of economy geography is expanding its dimensions and developing relationship with other branch of geography as well as other disciplines.

The techniques of land use study includes field investigation prepare of field maps design of sampling and presentation of data by different maps and diagrams and methods suggested by scholars in diverse fields of land use study.

RESEARCH DESIGN :

The physical and socio-economic environment of Dhanbad District is characterized by diversity. This is reflected in variation in climate, soil, land ownership, literacy and caste structure. Variation are also evident in the level of technology applied in agriculture in a varied an complex physio-socio-economic condition which is a function of the above factors is found to vary. This intra-regional imbalance has in the land use pattern has been studied into two levels the anchal level (meso level) and the village level (micro level).

For assessment of pattern in land use study at the micro level, representative villages from all over the area were selected on the basis of stratified purposive sampling. Using the anchal (Block) as a base, one village is selected from each anchal. Eight villages are selected from here. Selection of villages is carried out keeping in mind the following points.

I. To undertake a comprehensive study of the physical and cultural factors effecting land utilization.

II. To work out clear picture of the existing and past land use on the micro pattern on the basis of the details field survey, collection of data and cartographic representation.

III. To undertake a systematic examination and scientific analysis and interpretation of the exiting and past land use and thus to evaluate the various factors which have determined the very complex and indicate patterns of land use.

IV. To adjudge the potential land resources and work out land capability classification keeping in view the degree of limitations in the use of different categories of land, fertility status of soils, their slope, drainage and other proprieties, present and past land use etc and find out whether the land resources are properly utilized.

V. To suggest the actual land use planning for the future, i.e. the optimal use of the various category of land according to their capability and suitability in the national interest and to suggest proper steps to eradicate the handicaps and problems the hamper such as utilization.

VI. On the basis of the above study, to suggest the method and techniques of land use analysis, most suited to our condition.

The objective of the study is to undertake a detailed intensive land use survey of this Dhanbad District and to analyze and interpret it scientifically with a view to exploring the potential land resource of the region and suggesting on optimum use for the future in the national interest and thus converting this region of backward economic into a region of sound and flouring economic and also suggesting the suitable methods and techniques of land use analysis for such other area in the economy.

SOURCES OF DATA :

Materials have been collected from numerous sources and they include both published and unpublished, maps and literature, census data, unpublished data from official source and personal field investigation. The district settlement reports enabled minute statistics. The continues khatian provided every details of the land use in the village where as khasara revealed yearly record of land use plot by plot, statement or jinswar could be available of different seasons like Bhadai, Aghani, Ravi and Garma. Certainly some of these records are not faultless for they are collected by village karamchari, who is usually not sincere in abstaining correct data due to pressure of work and lack of stick checking and supervision. The reliability of data depends upon the efficiency of conscience of village official.

The published and unpublished census data were collected from census office and the published district census hand book of Dhanbad district of the year 1961, 1971, 1981, 2001 and 2011. The land use data, agriculture statistics of crops, harvest and average yields at block level were obtained from the Directorate of statistics and evaluation Patna. Various other data of sample villages were obtained from the karamchari, V.L.W, Circle officers and Mukhias.

In course of the present work the investigator had to visit the areas several times, especially, at time of different agricultural seasons for the collection, Puja vacation, X-mas and Holi were devoted in the field work.

Several maps were collected for the village map on the scale of 16"=1 mile, where every plot is numbered and demarcated. It has, therefore been easy to record land use of every plot. The base map is based on the scale of 1 cm=2 k.m.

METHODOLOGY :

Since the study area is large enough for detailed the use survey and analysis it has not been possible to study on village label data. Primarily the study is based on block level published and unpublished data and detailed of study of revenue village have been made on the sample survey of villages, one village from each block. In the selection of sample village, great care has been taken because the selection of sample village is considered an important factor determining the accuracy and usefulness of the result. The objective of sampling is to select a small numbers of individual cases which will present a reasonally true picture of the entire area. Normally sample survey is done on the basis of certain percentage of total numbers of the units in the area but only

8 villages have been selected from different areas represent all the problem and prospect of agricultural operation and land utilization for the entire area have been taken for the study on the basis of several varying physical and socio-economic condition.

While selection 8 villages for case study the consideration of the slope of the surface, potentialities of the land, availability of the irrigation water, means of communication and socio economic condition of the people were also given due consideration. Certainly in this work a purposive random sample method have been applied. First of all it was decided to select one village from each block. Then while selecting one village from each block, the consideration of slope, soil fertility, irrigation facility, and other determinates of land use were consideration.

Further all source of published and unpublished data were processed and then data were represented by suitable maps and diagrams, several choropleth maps, pie-diagrams and graphs have been constructed. Some colored maps of district label have been given for being attractive viz relief map, transport map and administrative map etc.

PROBLEM FACED :

While doing field work, collecting information from government offices and analyzing the data, investigator had to face a lot of problems. As regards problems faced in course of field work it should be noted that selection of study area sometimes poses problems. The selection of large area becomes difficult for intensive field observation and collecting first hand information. In fact intensive work is only possible for very small area like few villages or Gram Panchayat etc. Certainly the study of district level in not possible on village label data. No doubt, selection of sample village and their analysis to some extent solve the problem of study of large area, but it would be as accurate as detailed study of the smaller unit.

The study of large area needs more manpower for field work, data processing and their analysis which one investigator cannot complete, within limited time with meagre resource, financial help only the study of small area would be more appropriate.

Field work also involves a lot of problems which an investigator has to face. Unless he has some acquaintance or familiarity with the local people, collection of data from rural areas become difficult. Normally cultivators or lower level government authorities refuse to provide necessary maps, unpublished data and sometimes even hesitate to entertain. Luckily, the investigator comes from the same district and he has several known persons in the district, the difficulty of the field work and collection of data were not so difficult.

Frequent changes of administrative unites also poses problems for land use studies. Smaller administrative unit like police station, block, subdivision and even district boundary have changed and collection of data these level has become a problem, the comparative study of different censuses years becomes rather very difficult. Even the study of land use changes poses a lot of problems because of changes in administrative set up.

Published and unpublished land use data or crops data available from different sources differ and investigator is put into the problems, about their geniuses. Investigator struggle a lot to detect to fault of the information and furnish correct information in the study.

While in the process of collection of data and processing and analysis the need of aerial photograph highly realized. Because aerial photograph of different period, different agricultural seasons of great help in the study restricted availability of aerials photograph become hindrance in the proper and authentic land use studies.

SAMPLING TECHNIQUES :

In order to assess the land resources and to prepare a land use planning, it is necessary to have an intensive study of the land use of each village situated in the study area. Dhanbad district consists of 1348 revenue villages. This is why it is practically different, if not impossible, to record the existing use of every parcel of the land of all the 1348 villages. Specially when cropping pattern changes with seasons and from year to year. As the size of farm is very small it is very difficult to record the nature of land use. Such as large survey would involve mobilization of vast resources of manpower and organization,[45] as was done in the first land use survey of Great Britain 1930-45. That is beyond the scope of alone research worker to have detailed survey of all the villages of the region. Therefore in this chapter the only alternative is to select some representative villages for detailed study. Sample survey is recognized as the best method of selecting representative units all over the world. In India also the national sample survey organization has been set up with this objective in view. This organization has carried out a dozen sample surveys so far on which the whole scheme of the country planning and development is based. The sample technique seems to be the best for obtaining factual information in regard to the existing pattern and problems of land use.[46] The desired results are obtained by studying in detail only representative portion of the study area. There is no denying the fact that sampling cannot be substituted for total survey. But it is undoubtedly of considerable help. The sample may reveal the land use problems of the entire study area particularly if the area is homogenous in regard[47] to its Physical and socio economical conditions. Sample survey; therefore, gives good results at low cost and within minimum time. Sometimes it gives better results than the total survey.

There are many techniques of sampling and any one of them may be used depending upon the objective of the study. For the purpose of land use study three sampling technique are important, simple random or probability sampling, systematic or stratified sampling and purposive sampling.

In simple random sampling, which is regarded as the most common technique, the selection of sample unit is mechanically done independent of human judgment. In this technique efforts should be made to consider all the possible choice, so that every item shall have the same probability of being included. The stratified sampling is more representative than the former because the entire area from which the sample is drawn is divided into a number of strata and then random[48] samples are selected from within each stratum.

The arrangement of strata is such that each of them differs from the other as widely as possible but the population within each stratum is as homogenous as[49] possible. Purposive sampling on the other hand is a method of stratified, sampling in which the selection within the stratum is non-random. In this method after deciding the general break down of the samples on the scientific lines, the researcher is at liberty to choose the actual sample units. Although this method gives the impression of being less sound as compared to random sampling, it can be made highly[50] reliable with adequate safeguard.

The selection of the villages on random basis for land use studies may lead to a less representative picture of the area than that obtained from purposive sampling, because the human factors effecting land use usually vary in respect of farm management, accessibility etc. in each stratum. The systematic purposive cluster[51] sampling is most suitable for land use survey in India. Therefore, the systematic purposive sampling technique has been adopted in the present work.

The district of Dhanbad is divided into 8 Anchals. These Anchals vary from one another in physiography, drainage, soil, vegetation and general land use. Therefore, these Anchals have been recognized as stratified units. One village has been selected from each of the Anchals. Thus, altogether 8 villages have been studied in detail. The details regarding

Fig. 1.1

Source : The Map Prepared by Author

the villages are shown in table No.1.1

A glance of the distribution (figure No. 1.1) of the villages show that they are wide spared throughout the length and breadth of the district. As regards location of the villages they are situated between 23°40' to 24°0' north latitude and 86°15' east to 86°50' east longitude. Some villages are situated near national highway and state highway. They are easily accessible. These villages are Tetulia, Sialgudri and Saldaha. The village Rohraband is just along Jharia, Sindri road. The Paharpur village is located just by the Govindpur, Baliapur road. The village Bangaro and Behachia is located near hilly portion of the district on Rajganj Maniadih road. The village Kharni is located nearby Rajganj Katras road. All these village are easily accessible by road.

Table - 1.1
SAMPLE VILLAGES OF DHANBAD DISTRICT

Sl. No.	Name of the Village	Name of the Anchal	Population 2011	Area in Acres	Thana Number
1	Bangaro	Tundi	654	541.93	73
2	Behachia	Govindpur	788	516.96	1
3	Tetulia	Nirsa	905	485.37	133
4	Paharpur	Baliapur	1819	529.77	30
5	Saldaha	Baghmara	334	237.09	200
6	Kharni	Topchanchi	951	380.90	153
7	Sialgudri	Dhanbad	5450	253.38	81
8	Rohraband	Jharia	8248	1152.91	169

REFERENCES

1. Foreman, W. (1968) Geography and Planning, (London: Hutchinson University Library), p.74

2. Nanavati, M.B. (1957) Reading in Land Utilization, The Indian Society of Agricultural Economics, p. 2

3. Stamp, L. D. (1948) The Land of Britain and How it is used (London: Longman) pp.74-77

4. Stapledon, R. G. (1944) The land : Now and Tomorrow, LONDON Fahar and Fahar, p. 32

5. Singh, R. P. (1967) 'Concept of land use' Patna University Journal, vol. 22, No. 1, pp. 52-62

6. Klages, K. H. W. (1947) Ecological crop Geography, New York, p. 10

7. Stamp, L. D. (1962) The land of Britain : Its use and misuse, London

8. Stamp, L. D. (1962) Applied Geography; Bungay : Penguin book, Richard clay and Co. Ltd., pp. 48-50.

9. Kostrowiekai, J. (1968) Agricultural Typology, Agricultural Rationalization and Agricultural Development, 21st International Geographical Congress, Abstract of Papers, Calcutta.

10. .... (1945) Land Utilization in the District of 24 parganas, Bengal. B. C. law

11. Chatterjee, S. P. (1952) 'Land utilization survey of Howrah district', Geographical Review of India, vol. 14, no. 13.

12. Shafi, M. (1961) Land utilization in Eastern Uttar Pradesh, Aligarh Muslim University, Aligarh.

13. Shafi, M. (1966) 'Techniques of Rural land use planning with Reference to India', Geographer, vol. 13.

14. Shafi, M. (1972) Land use studies : A trend Report. A survey of research in Geography, Popular Prakashan, Bombay, p.19

15. Shafi, M. (1969) 'Can India support Five Times her population, Since Today', p. 24.

16. Ahmad, E. (1954) Geographical Essays on India, Patna, pp. 15-16.

17. Dayal, P. (1972) 'Land use Pattern in the Biharsharif area', Patna saran A district.

18. Sinha, R.N.P. (1965) 'Present land use pattern in canal irrigated area of Patna district'.

19. Amani, K. Z. (1968) 'Land utilization in village, Golagarhi'.

20. Ayyer, N.P. (1968) 'Crops Region of M.P. : A Study in methodology', Geographical review of India, vol.-xxx1, No. 1 , pp. 4-5

21. Sharma, H.S. (1972) 'Land capability classification of the lower Chambal valley', proceeding of symposiums on land use in Developing country. 21st international geographical congress, Aligarh.

22. Learmonth and (1968) 'Recent contribution to the Economic Bhat, L. S. Geography of India : some current preoccupation', Economic geography, vol. 44, No. 3, pp. 189-209

23. Singh, V. R. (1970) 'Land use pattern in Mirzapur and Environs'. Unpublished Ph.D. Thesis, Banaras Hindu University, Varanasi.

24. Das, K. N. (1969) 'Population and land used changes in the kosi region, Bihar' Unpublished Ph.D. Thesis, Bhagalpur University.

25. Roy, S. R. (1976) 'Land use and settlements in the District of Bhojpur and Rohtas, Bihar', unpublished Ph.D. Thesis, Bhagalpur university. 'Soil survey and Land use Analysis'

26. Rao, V. L. S. P. (1947) IndianGeographical journal, Vol. 22, No. 3

27. Lahiri, R. (1950) 'Land utilization in some village near Jasidih', Geographical Review of India vol.12, No.1

28. Karimi, S. M. (1949-50) 'Land Utilization in Dinapur Ghuspara' The Geographer, Aligarh Muslim University, Aligarh

29. Bharadwaj, O. P. (1960) 'Land Use in Bist Jalandhar Doab' National Geographical Journal of India

30. Rai, B. K. (1968) 'Measurement of Rural Land Use in Ajamgarh Middle Ganga Valley'

31. Mandal, B. (1969) 'Crop Combination Regions of North Bihar' The National Geographical Journal of India.

32. Sinha, B. N. (1968) 'Crop Combination Technique: A Search for an Ideal Tool' Deccan Geographer

33. Tripathi, V. B. (1967) 'Changing Pattern of Crop Land Use in The Lower Ganga Yamuna Doab'

34. Garg, S. P. (1968) 'Land Utilization in Saharanpur District' Unpublished Ph.D. Thesis Agra University.

35. Jha, B. N. (1980) 'Problems of Land Utilization, A Case Study of Kosi Region' New Delhi

36. Yadav, J.P.S. (1965) 'Crop Land Use Pattern in Rajasthan' Conference of Geographers, Bhilwara.

37. Priya, I. (1971) 'Land utilizations in Bhandro Block, Ranchi, unpublished Ph.D. thesis, Magadh University.

38. Prasad, M. (1977) 'The changing pattern of land use in Bodh Gaya block, (Gaya),' unpublished Ph.D. Thesis, Magadh university.

39. Ataullah, M. (1979) 'Urban land use in major town along the Ganga in Bihar', Unpublished Ph.D. Thesis, Patna University.

40. Mandal, R. B. (1982) 'Land utilization theory and practice' Unpublished Ph.D. Thesis, Patna University, Concept Publishing Co., New Delhi.

41. Govt. of India (1958) 'Agricultural Atlas of India'.

42. Chatterjee, S. P. (1957) 'National Atlas of India'.

43. Census of India (1961) States atlas of Andhra Pradesh, Bihar, Gujarat, U.P. (1961), Vol.-1, Part-1A (!) Levels of Regional Development in India.

44. Stamp, L. D. (1968) Land of British: its use and misuse London

45. Shafi, M. (1966) 'Technique of rural land use planning with reference to India' the geography, Vol. XIII, p. 16

46. Mohammad, Ali (1977) Food and Nutrition in India, New Delhi, p. 47

47. Kish, L. (1963) Survey sampling, New York, p.75

48. Shafi, M. Op cit. p. 47

19. Ibid p. 48

50. Noor, M. (1978) Agricultural land use in India, New Delhi, p. 47

2

Chapter Environmental Framework

Dhanbad district is a part of north Chotanagpur plateau, which have a little productive area in the north continuous ridge of Parasnath Range run for a long distance form the hilly portion of the district. In the south occupied by the great coal basins of Jharia. Density of population in the north is less than south mostly people in the north depend on agricultural, which grows mainly rice, maize, and vegetable. A person of southern portion is mainly concerned with trade and services. In this chapter a brief account of physiographic, drainage, climate, vegetation, soil, population and along with physical economic and social setting and occupational structure has been analysed separately before studying the land use changes.

LOCATION :

Dhanbad is an important district of Jharkhand. It is located in south-east part of the state of Jharkhand. It is a small district. It lies 23°37'3" to 24°40" N latitude and 86°6'30" E to 86°50" E longitudes. The district is bounded on the north by Jamtara and Giridih districts, on the east Burdwan district (West Bengal). On the west and south by Bokaro district. The river Barakar flows on the northern and eastern part of the district while the Damoder River flows on the southern and south eastern part of the study areas. Dhanbad town is the head quarter of the district. It covers an area of 2040 sq.kms with a density of 1316 person per sq km (2011) which shows a crow fly length about 75 kms from east to west and 45 kms from north to south. It has 8 development block consists of 1348 villages.

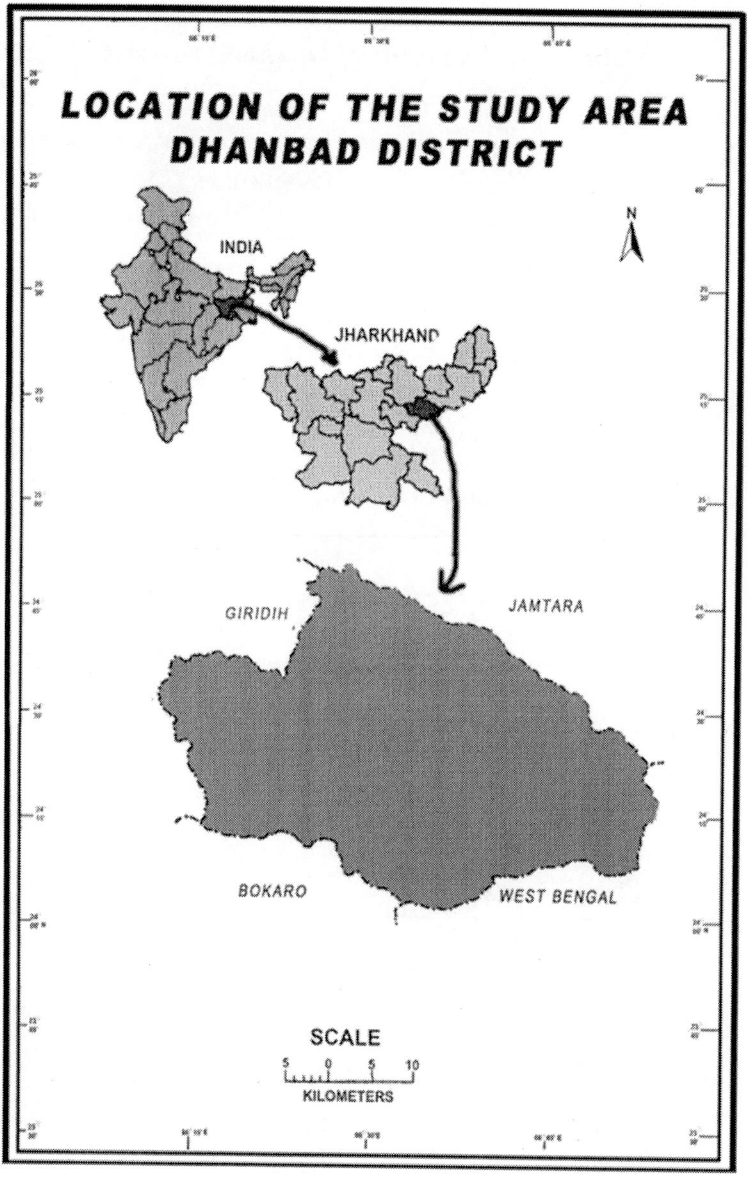

Fig. 2.1

Source : The Map Prepared by Author

Table - 2.1
Administrative units of Dhanbad District

1.	Subdivision	- 01	(Dhanbad)
2.	Development Blocks	- 08	(Dhanbad, Topchanchi, Baghmara, Jharia, Tundi, Govindpur, Nirsa, Baliapur)
3.	No. of police stations	- 26	
4.	No. of villages	- 1348	
5.	No. of Panchayats	- 256	

Source: District Census Hand Book of Dhanbad

Table - 2.2
Dhanbad Important Statistics

1.	Total Area	-	504094.97 acres or 2040 km^2
2.	Net sown area	-	103944 acres
3.	Total Population	-	26,84,487 (2011)
4.	Density per square km	-	1316
5.	Average Annual Rainfall	-	1307.9 mm
6.	Average Annual Temperature	-	25.5 c
7.	No. of Urban center	-	27
8.	Literacy Rate	-	75.71%
9.	Sex Ratio	-	909
10.	Population Growth Rate	-	11.91 (2001-2011)

BRIEF HISTORY :

Dhanbad was a part of Manbhum district of West Bengal. It became a part of Bihar after Bengal division and finally of the state of Jharkhand. That times Dhanbad was called "DHANBAID". It has a long story of development. It is believed that Mr. luby I.C.S wrote Dhanbaid to Dhanbad first time and took permission to drop the letter "I". Thus[1] Dhanbaid become Dhanbad from 1918. From 1 Nov. 1956 Dhanbad became a district. At that time it had an area of 2926 sq kms. In 1995 Dhanbad was divided into two parts

Dhanbad and Bokaro. During this division Chandan Kiyari block was given to Bokaro district. At present district has an area of 2040 sq Kms. Dhanbad is divided into eight development block. According to the census of 2011 the total population of the district is 2684487 persons.[2]

There is no authentic record to show how Dhanbad took

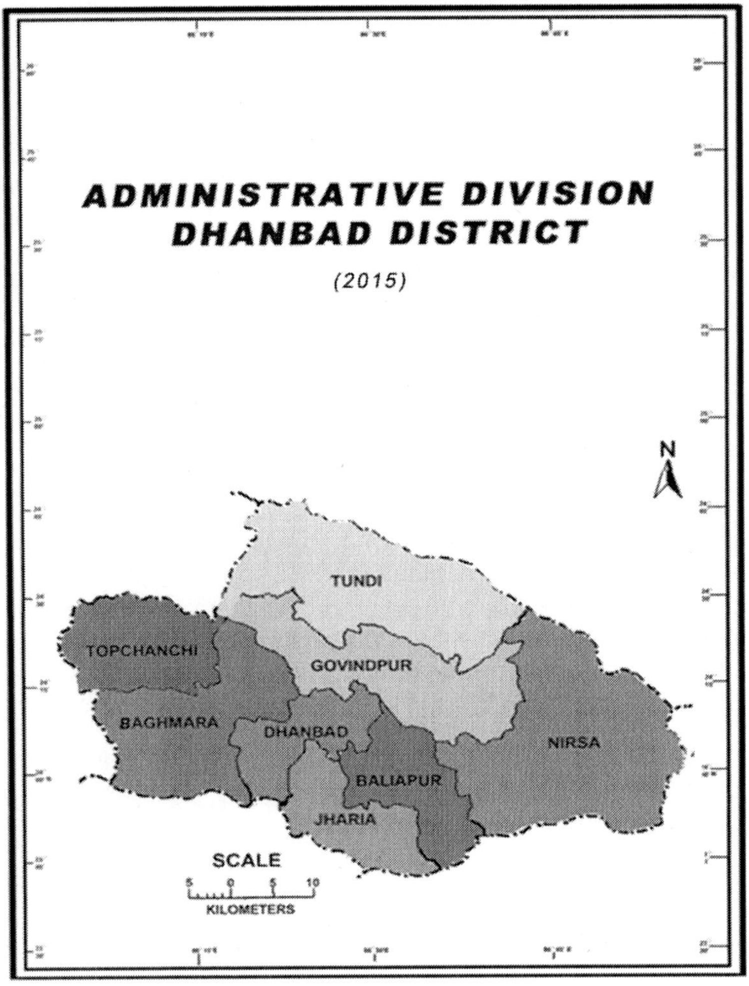

Fig. 2.2

Source : The Map Prepared by Author

its name. One of the popular opinions among the public is that this area was famous for the growth of "BAID DHAN" or Baid paddy. There are two kinds of paddy in the district. One is called "Baid" which ripens in kartik (October-November). Another far-fetched theory is that the name of Dhanbad is derived from Dhan, a kolarien tribe that lived in this area. Baid has perhaps been taken from the Urdu word "Abad" meaning occupied Dhanbad may mean the palace populated by "Dhan" a kolarian tribe or occupied by 'Dhan' paddy in general.

The early history of the greater part of the Chotanagpur plateau is shrouded in mystery and that of Dhanbad district is particularly so. Details of even later periods are difficult to trace as the present district formed but a small and insignificant part of Manbhum. In the settlement report for Manbhum (1928) it is stated that no rock inscriptions copper plates or old coins were discovered and not a single documents of copper plates or old coins were discovered and not a single document of copper plates for palm leaf was found in course of the survey and settlement operation. The oldest authentic documents produced were all on paper and barely even a hundred year old.

In view of these circumstances the District Gazetteer for Dhanbad (1964) merely reproduces the chapter on history in the settlement report of 1928 which pertain to the entire district of Manbhum.

The earliest civilization in the district was that of the Jains who are cribbed by Colonel Dalton to 500 or 600 B.C. The Saraks who were the descendants of these Jain, were found in considerable number in the eastern part of district was as late as in 1928. The Jain civilization was apparently superseded by the Brahmins before the seventh century A.D. The travel accounts of Hieun Tsang in seventh century A.D make mention of a powerful kingdom ruled by Sasanka, a perrecutor of the Buddhists, which extended from Midnapur on the east to Surguja on the west. Accordingly to one view it is held that Mundari tribes occupied the area south of Bihar in Pliny's time. Brahmanical civilization was at its prime in the tenth century. It was destroyed completely by the Bhumij, a Mundari tribe, it is started that the area occupied by the latter mainly extended to south of the river Kasai in the area that is now largely contained is Purulia district of west Bengal. It is belived that during Akbar's rein in circa 1590, Raja Man Singh led his troops from Bhagalpur to Midnapur through Jharkhand and he might have passed through portion of Dhanbad.

PHYSICAL BACKGROUND :

Geology

Study of geology of a particular region is very much important. From geological point of view Dhanbad district can be divided into two equal parts northern half and southern half. The northern half which consists entirely of ancient crystalline rocks, and the southern half which is occupied by the great basin of Jharia in the west and the Barakar portion of the Raniganj coal basin in the east with intervening area of crystalline rocks.

The oldest geological formation of Dhanbad distict are composed of crystalline metamorph rocks which belongs to Dharwar systems (Archeans) in Indian statigraphy. In these information of both sedimentary and igneous origin rocks are found. The sedimentary rocks were originally deposited as sandy, clayey and calcareous sediments more or less impure.

These sediments are consolidated as sand stones, shales and lime stones of different composition and them are subjected to regional metamorphism and changed in to quartzite's micaceous schist, crystalline, limestone, cale-silicate, granules and calc-gneisses and in same cases to infobolities and hornblende schists. The last two rocks types may impact the contemporaneous lava flows or sills subsequently metamorphosed. All has happened in Dharwar period.

The above metamorphosed sedimentary rocks were then intruded by magma of both basic and acid composition, possibly in the earlier post Dharwar times. The basic magma was of dolerite and nortic composition and is now respected by intrusive dykes and masses of met dolerite, meteorites and some of the epidiorites. These rock types usually occur as prominent hillocks in the area. The acid magma which intruded later than the basic magma had given rise to the granites, pegmatites, aplites and some quartz veins of the area.

In the later post Dharwar time the metamorphosed sedimentary rocks and the igneous intrusive were subjected to the injection metamorphism due to which the rocks types were permitted isolated, felspathised, homogenized and granitised in to streaky, angen and injection gneisses.

The ancient rocks of the Dhanbad district is of the lower Gondwana group of sediments strata including the coal bearing beds which laid down. River deposited in slowly sinking faulted basin in several thousand feet of river and stream. At the commencement of the lower Gondwana period since 300 million years ago, there was a glacial age in India as it evident from the glaciated boulders deposited at the bottom of the Gondwana system of rocks and the presence of under composed feldspar grains in the associated sand stones above them. The climate rapidly became much warmer and gave rise to the super abundant growth of vegetation which supplied the materials for the formation of coal in the succeeding series of rocks in the Jharia coal fields and in the Barakar portion of the Raniganj coal field. The main axis of the synclinal Jharia coal basin runs west of north, south of east and is pitching gently towards west as can be seen by the dips of the Raniganj strata along the Jamunia stream. This line of axis is neither straight nor curved in a simple manner but it is itself corrugated due to the forces probably coming from east to west.

The recent formation which from superficial deposit, the alluvial sand of river and stream, and the residual soil derived

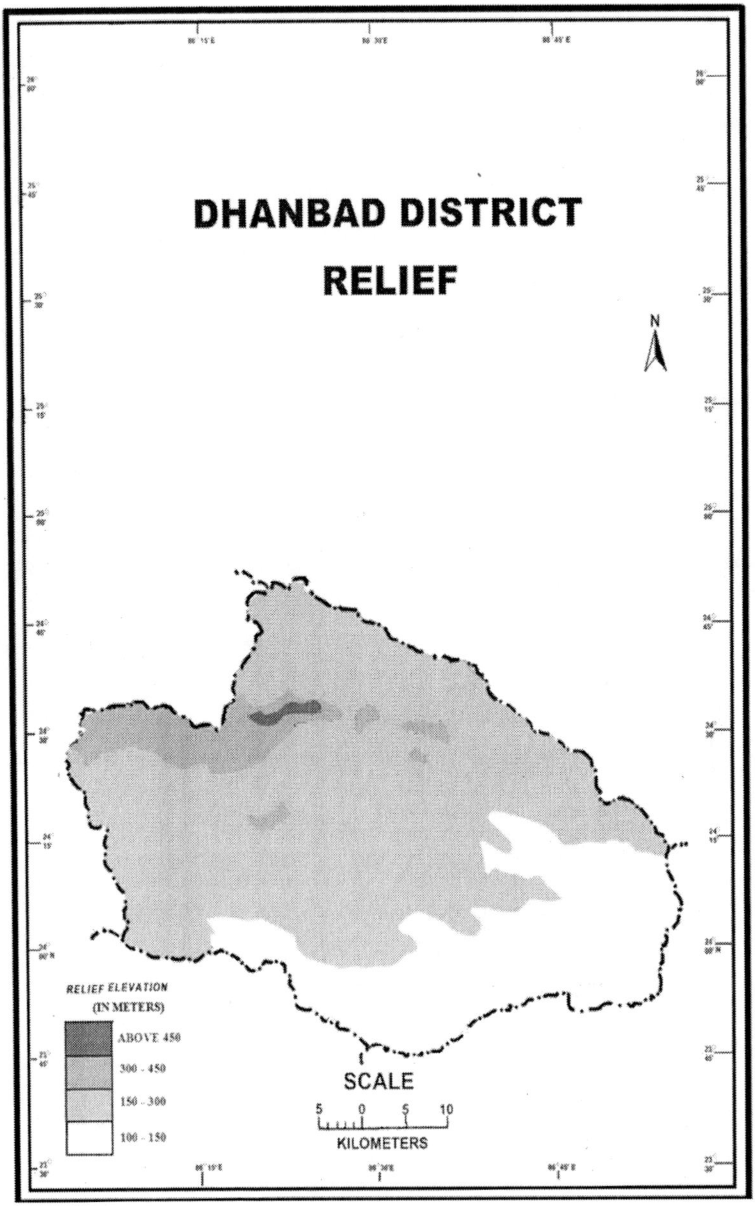

Fig. 2.3

Source : The Map Prepared by Author

from the weathering of bed rocks. Thick deposited of recent alluvial sand occur adjacent to the banks of Damodar and Barakar river and the Jamunia, Khudia and Pusai streams. The Damodar sand is being used to places for stowing the goof areas in the important coal basins of the Jharia coal field. In addition to recent alluvium. There are patches of older alluvium on the banks of the Damodar and Barakar rivers, lying either well above the present high flood level of the rivers or in channels. Now more or less abandoned due to the shifting of the river course.

DRAINAGE :

The drainage line holds particular significance in the region. It has very close relation with the structure of the region. Although the drainage of a particular area depends upon the amount of rain, it receives during the rainy season. All rivers are rain fed and seasonal in character.

The general slope of the land is towards south and south east. The Damodar River is itself flowing from west to east, along the southern boundary fault of the Jharia coal field. All the rivers which intersect or take their rise in it have the usual characteristics of hill streams. All rivers remain dry during the greater part of the cold season and in the hot season. They are not navigable during any part of the year with the single exception of the Damodar River and subject to sudden and violent flood which are usually of very short duration. Except where they run over exposed rock their beds are usually deep in gravel and sand the banks are abrupt and broken into deep cuts wherever the drainage the surrounding country finds its ways to the level of the stream. The northern most river in Dhanbad is the Barakar which forms the northern, north eastern and eastern boundary of the district and runs for about 77.23 Kms with in this district. The Barakar is the most important tributary of the Damodar. It meets the Damodar River near Chirkunda.

Damodar is the most important river of Dhanbad, which forms the southern south eastern boundary of the district and run for about 112.63 km within this district. The Damodar rises in Chhotanagpur hills about 60.96 meters above sea level flowing north east till it meets the Barakar. After a run of about 563 km the river fall in to Hoogly. The Damodar runs about 290 km in Jharkhand before entering West Bengal.

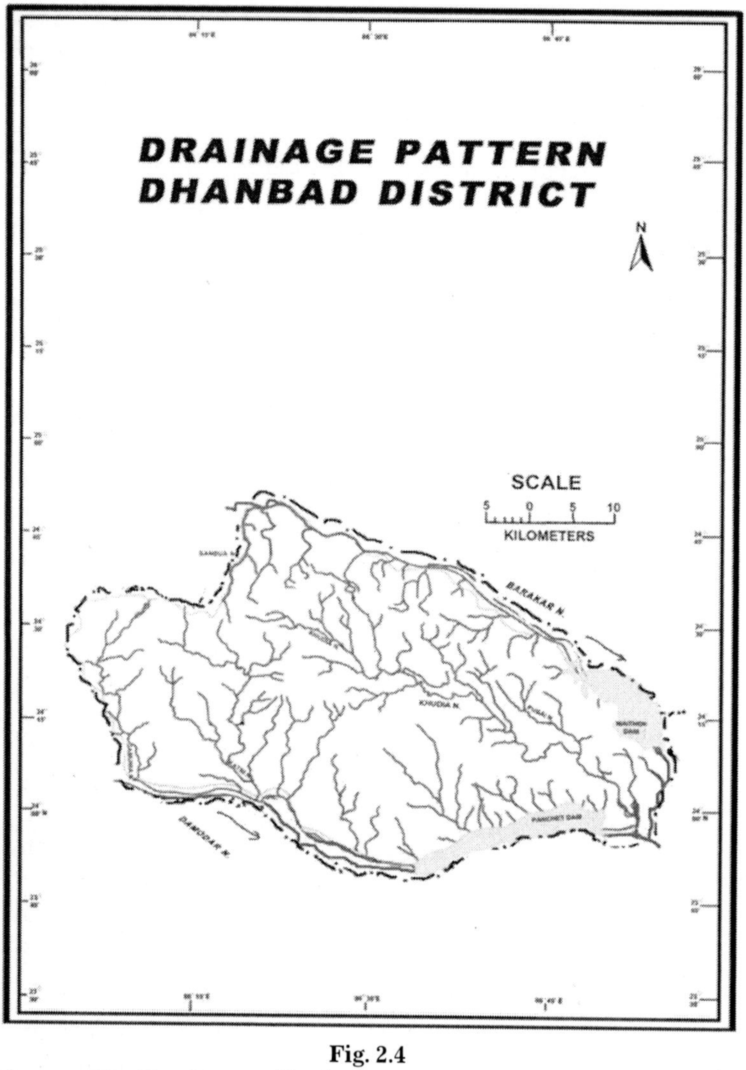

Fig. 2.4

Source : The Map Prepared by Author

Besides the above rivers there are other small rivers also. They are as follows:

1. The Jamunia river - It is main tributary of the Damodar River and flows north. It marks the north western boundary of Dhanbad district. It starts from Parasnath hills after running about 51 km. It meets the Damodar River near Gomoh.

2. The Khudia River - The second important tributary of the Damodar is Khudia River. It starts from the middle of the village Parasbani and Asanbani and end at east from and south side of the district near Chanch. It runs for about 46 km in the district. Khudia River flowing towards south east wards, north of Govindpur with its tributary Pusai. Due the prolonged denudation to which this region has been subjected, there is not much correspondence between the structural features and the direction of drainage in case of river other than the Damodar.

3. The Katri River - Another important tributary of Damodar is Katri. It flows from north. It starts near Tilaiya village and runs about 64 km it before meets the Damodar.

PHYSICAL DIVISION :

The region is a part of Chotanagpur plateau. The topography of the southern half of the area is undulating and dull with very few conspicuous feature. The northern portion is how ever characterized by greater variation in relief. The upland lies mainly in the northern portion of district. There are no large stretches of what may be called as plain in the district. A part from the several low ridges which occur north of the Grand chord Railway line between Pradhan Khanta and Gomoh station. The highest peak of Parasnath 1365.50 meters above sea level sends out its spurs up to north western corner of the district where the famous Topchanchi reservoir is built to supply drinking water to Jharia coalfield. The more or less continuous ridges of the Parasnath range run for a long distance eastward from the place and from the northern hilly portion of the district. These ridges vary in altitude from 365 meter to 381 meter above the sea level. Dhangi hill run from Pradhan Khanta to Govindpur in this district. The highest peak in these hill is at Dhangi is 385.572 meters high.

Most of the hills and ridges of the area owe their preservation due to the greater hardness of the rock which constitutes them compared to that of the surrounding rocks. The common types rocks which give rise to outstanding hills are the metamorphic rock like epidermises, amphibolites, met dolerites and meteorite. Some of these hills may thus be regarded as igneous in origin. A few ridges are composed of quartzite, granulites and micaceous, schist's and gneisses and they may regarded as relict type of ridges, which are suffered less erosion than the surrounding area. The views of white quartz, which have suffered less erosion than the surrounding area. The view of the white quartz which are so common in the metamorphic terrain of the district also from low ridges due to their resistance to denudation. These ridges may how ever be regarded as tectonic in origin as they usually indicate fault zones.

Broadly Dhanbad is divided into three natural regions. These are:-

I. The range of ridges in the northern and north western region covering an area of about 136 square kilometres.

II. The upland containing coal field having approximately an area of 321 square kilometres in the southern and eastern parts.

III. The series of up lands and intervening hollow with isolated bare ridges of varying between them.

CLIMATE :

Among all the components of the physical environment climate is the foremost which affect land use. The characteristics of the soil are largely the product of present and past climates and the vegetation that has flourished in them. Relief is too expressed in so small degree in climatic variation as well the farm operations such as ploughing sowing of seeds, transplantation, irrigation, manuring, harvesting threshing, winnowing, sugarcane crushing are governed by the favourable weather conditions. The climate of the Dhanbad region is essentially transitional between more humid. Bengal in the east and the comparatively dry

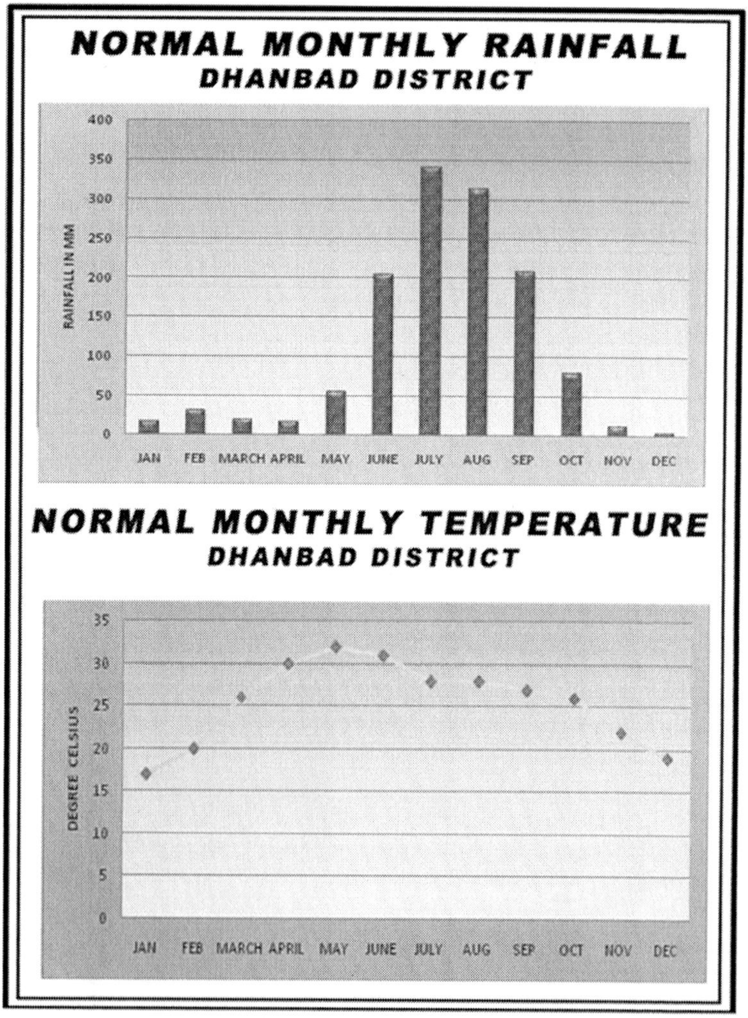

Fig. 2.5

Source : The Map Prepared by Author

Uttar Pradesh in the west. It experiences sub tropical monsoon type of climate. Due to its location the climate of the region is controlled by few factors. Firstly, it is located between the wet and humid climate of Bengal and Assam in the east and relatively dry climate of the west. The region is open in the east and west for the free entrance of oceanic monsoon winds from the east and continental dry wind from the west. The location of Himalayan mountain ranges in the north influences the direction of the winds and also the temperature of the area. It hinders the central Asiatic cold wind from entering the area. As such the temperature does not fall as central Asia. The area is located north of the tropic of cancer, so it experiences tropical type of climate. The nearness of the Bay of Bengal is another factor due to which the temperature of the region is not as high or low as that in the central and northern part of the area. The region is also influenced by the movement of sun from tropic of cancer to that of Capricorn. This change gives birth to the monsoon, which influences the whole of India including this region.

On the basis of temperature, rainfall and of atmospheric disturbances, three distinct season with two well marked transitions are found here. These seasons are as follows.

I. The cold weather season - November to February

II. The hot weather season - March to mid June.

III. The Rainy season - Mid June to October.

The data regarding monthly rainfall and temperature of the district have been taken from the average of 50 years are given as follow in table-2.3.

Table - 2.3
Average of 50 years temperature and rainfall

Month	Temperature in °C	Rainfall (in cm)
January	17.70	15.10
February	20.40	26.00
March	26.00	19.70

April	30.20	17.60
May	32.00	51.30
June	30.60	193.70
July	28.00	343.90
August	27.60	320.10
September	27.40	210.70
October	25.80	93.20
November	21.60	10.90
December	18.50	4.10
Average	**25.48**	**109.69**

Source: D.V.C office, Maithon

I. THE COLD WEATHER SEASON :-

The area enjoys a long cold weather, which commences early in November and ends in February. Clear sky, fine weather, low humidity and the large diurnal variations of temperature are the characteristic features of the seasons.

The fall of temperature starts with the rainy season and continues till the end of winter season with the movement of the sun towards the tropic of Capricorn both day and night temperature decreases fairly and rapidly with the advance of the season. January is the coldest month with mean daily temperature at 16.5°C. Sometimes the temperature goes down to 4°C to 5°C above the freezing point of water. The temperature of November and December are 22°C and 17°C respectively. February experiences a temperature of 19°C.

This season experiences continental wind, which is devoid of moisture this, is why the region does not receive significant amount of rain during the season. November and December are generally devoid of rains whereas January and February receive some rainfall due to eastward movement of temperature cyclone in January rainfall is more marked in the intensity than the February. The variability of rain fall during this season is also well marked. The rainfall through small in absolute amount is of considerable agricultural significant as the character of Rabi harvest depends upon it.

The air in the early hours of the morning is saturated and there is copious dewfall, which is highly beneficial to winter crops. By the end of February and the beginning of March the temperature tends to increase.

II. THE HOT WEATHER SEASON :

The hot weather season starts in March and reaches its peak in mid June. It is a hot dry period when a westerly wind prevails. This is marked by severe dry heat. In this season the area is very unpleasant where nights are uncomfortable. Bad days are marked by Loo[3], the dry hot westerly. The temperature rises to more than 45°C as the summer season advances. It makes people to confine themselves in their houses from 10 am to 4 pm. The temperature begins to rise in late February. By March it reaches 25°C. In March the days become warmer while the night continues to be cool. With migration of the sun northward both day and night temperature begins to increase rapidly after middle of the march till May, which is the hottest month of the year. April and may experiences 30°C and 31°C of temperature respectively the period of maximum temperature stretches up to the middle of June when with the onset of monsoon. There is a sharp drop in the day temperature. June experiences an average of 42.5°C temp. The month of June is in the grip of low pressure. The gently blowing westerly breeze of the early morning changes into high speed in the high speed in the afternoon. Sometimes the high speed hot winds Influence on the vegetation. Land surface lies parched and bare of vegetation. The hot weather season is on the whole intensely hot and confines to blow till late evening. It is locally Loo. It causes sunstroke and even leads to death. The excessive heat has a desiccating dry which make this a period of compulsory fallow over the greater part of the land.

The greater part of the westerly is dry and devoid of moisture. Sometimes it is accompanied by humidity and gives rise to heavy rain with storm. This rain is locally called Mango shower or pre monsoon rain. This rain is very significant for land use as it is followed by the start of the Bhadai cultivation.

III. THE RAINY SEASON :

With the advent of monsoon in June the temperature is bought down considerably. Normal year the monsoon breaks around the 12 June lasts till October. But sometimes it breaks either earlier or late. The late arrival of monsoon cause great harm to agriculture. Any how the onset of monsoon the weather changes completely and the restless night and days are converted into comfortable into comfortable one. People are full of joys and brisk. Agricultural operations starts and dried parched land are filled with water. The hot scorching weather becomes sultry with the rise in the relative humidity. During this period Cyclonic storms originating at the head of the Bay of Bengal also come with copious rainfall associated with thunder. With the arrival of monsoon in mid June the temperature begins to fall considerably while June experiences 42.5°c of temperature. July, August and September have an average temperature of 32°c. It is significant to note that days and nights having no rain experiences very high temperature. But due to rise in humidity it is scorching rather muggy. After June the rainfall goes on increasing till august which experiences the highest rainfall of the season. After august it begins to fall again. On the whole the region receives about 80% of its rainfall during the season. The average annual rainfall of the region is 1307.9 mm. June receives on an average 205.3 mm of rainfall, July and august receive 341.5 mm and 313.3 mm of rain respectively. In September it is brought down considerably when 209.4 mm of rainfall is received on an average. By October it begins to fade away and this month receives only 80.1 mm rainfall.

Table - 2.4

Monthly Rainfall in the District of Dhanbad (Millimeter)

Month	Normal	2009	2010	2011	2012	2013	2014	2015
January	16.7	0.0	0.0	1.7	14.6	0.0	11.8	15.8
February	31.3	0.0	7.5	0.0	9.1	22.5	0.0	1.2
March	19.5	1.4	0.0	3.3	0.7	0.0	28.4	4.4
April	17.1	1.8	5.8	17.8	7.6	30.0	1.1	57.9

May	56.4	77.4	47.9	3.3	18.9	177.4	88.5	23.3
June	205.3	56.3	15.3	387.9	83.2	225.0	225.01	62.4
July	341.5	237.3	21.5	179.3	346.2	168.1	346.2	346.0
August	313.3	311.4	237.8	491.0	276.4	268.9	311.5	141.6
September	209.4	303.8	30.6	319.1	326.9	191.6	303.8	89.6
October	80.1	83.5	71.7	20.1	12.6	300.7	71.7	8.7
November	12.7	8.9	0.0	0.0	62.5	0.0	0.0	0.0
December	4.6	1.6	14.1	0.0	10.6	0.0	0.0	0.0
Total	**1307.9**	**1083.4**	**452.0**	**1423.5**	**1169.3**	**1384.2**	**1388.0**	**850.9**

Source: D.V.C office, Maithon

The rainy season has a great bearing on the land use of the region under study. Before it the whole agricultural land look like fallow land where nothing is grown except around towns where market gardening is done and also where irrigation facilities enable the cultivation of vegetable. But the monsoon break the agricultural land is given to cultivate of different types of crops.

REGIONAL VARIATION IN RAINFALL :

In addition to the seasonal variation in the rainfall. There is a variation in the amount of the rainfall received in different parts of region. The location of hilly areas in the northern portion receives relatively more rainfall than remaining area. Baghmara, Dhanbad and Govindpur Anchal receive above 1300 mm rainfall. Tundi and Topchanchi also receive rainfall between 1300 mm to 1400 mm. only Nirsa receive rainfall less than 1100mm.

The area having relatively more rainfall produce paddy, maize and Ragi in the Kharif season and wheat, vegetable in the Rabi season. The production of Rabi crops is based on irrigation while Kharif depends mainly on rainfall. However, agriculture is largely at the mercy of erratic rain. The retreating monsoon helps in maturation of winter paddy and provides moisture for Rabi crops. Failure of Hathiya rain often leads to drought and crop failure. The rainfall pattern indicates that the Hathiya rain fall once in three year in the region, once in four year in the northern part of Bihar and once in five years in the Chhotanagpur plateau, which partially accounts for the failure of agriculture in the region. The inadequacies and uncertainty of rainfall make Jharkhand's agriculture a short of gamble on the monsoon.

Table - 2.5
Maximum and Minimum Temperature by month in the district of Dhanbad

(Degree Celsius)

Month	2010		2011		2012		2013	
	Min	Max	Min	Max	Min	Max	Min	Max
January	NA	NA	27.50	6.80	34.10	NA	NA	NA
February	NA	NA	33.60	10.40	27.00	30.00	30.00	23.30
March	NA	NA	39.20	13.00	39.20	34.90	34.90	24.00
April	NA	NA	40.30	18.60	42.30	37.80	37.80	32.00
May	41.57	23.58	43.50	20.90	43.50	43.50	43.50	34.20
June	42.50	23.88	40.80	21.40	36.90	36.90	36.90	33.20
July	36.29	24.56	36.30	23.00	35.40	35.40	35.40	26.10
August	37.07	24.27	36.70	23.20	35.90	35.90	35.90	28.30
September	33.70	22.00	36.20	23.20	34.00	34.00	34.00	30.90
October	34.60	16.70	34.50	15.40	31.20	31.20	31.20	15.70
November	32.30	12.68	30.30	13.20	31.30		31.30	11.10
December	27.20	7.90	28.50	6.80	27.30		27.30	9.60

Source: ISM Centre Dhanbad

It will be seen from table 2.6 that variability of rainfall during the monsoon periods is high. Rainfall is maximum in July and decreased gradually in the month on October, November and December. In winter season it is below 50mm. The high variability of rainfall during monsoon is due to the variation of the timing of onset of the monsoon. The above analysis shows that variability is higher in July. The rainfall in area is unreliable during the month when its regularly is most needed. Such uncertainties in rainfall especially in the area where agriculture is reduced top mere gamble with monsoon due to the lack of adequate irrigation facility have an adverse effect on agriculture.

Table - 2.6
Mean monthly rainfall (in mm) 2015

Name of Block	Jan	Feb	March	April	May	June	July	August	Sept	October	Nov	Dec	Total
Nirsa	7.5	39.1	53.9	61.0	218.7	95.9	294.1	171.0	59.9	78.0	11.4	6.2	1096.7
Tundi	2.4	55.6	35.0	58.9	220.2	262.3	262.3	237.5	190.4	95.2	6.4	5.2	1431.4
Govindpur	8.6	00	91.4	61.5	208.3	203.2	328.4	210.6	147.2	105.3	18.5	10.4	1993.4
Baliapur	2.0	00	5.2	50.5	192.3	185.5	345.3	202.2	307.2	160.01	1.5	7.3	1458.5
Dhanbad	1.5	24.01	78.7	50.6	192.3	332.7	345.2	202.3	328.2	108.5	18.6	6.4	1688.6
Jharia	3.0	25.05	21.8	50.5	192.5	205.2	345.4	292.2	179.3	21.6	5.6	6.5	1358.1
Baghmara	00	32.02	61.1	29.8	209.3	373.1	342.9	225.0	376.5	68.2	3.2	8.4	1729.3
Topchanchi	7.5	4.08	47.8	59.7	209.4	221.6	332.9	220.1	160.9	74.6	3.3	3.8	1344.7

Source: D.V.C office, Maithon

NATURAL VEGETATION :

The area under forest is very small though there is a large area lying barren. Which were sometimes covered by dense forest. The forest of this district is mostly in Tundi Topchanchi and Nirsa area, which are to the northern part of the district. Formerly when forest covered most of the land surface and clearance of jungle was necessary or laudatory, nobody cared as to what happened to the forest, cutting went on without let or hindrance. As the population grew more cutting took place and denudation was the result. Some of the important trees in the forest are Sal, Sisam, Siris, Palas, Mahua, Jamun and Simal. Bamboos are also Common. The names of a few of the common species generally occur in the forest of Dhanbad district are Gamhar, Karam, Karanj, Doka, Khanjoor, Koosum, etc. The total forest area of Dhanbad district is 170 sq .km or 56454 acres.

As has been said in the foregoing paragraph, large stretches of land, which once were covered with fine forests, have since decades been completely denuded of all vegetation and standout as desolate brown waste on which not even grass can grow, Large parts of such waste have been gullied and deep ravines formed therein. For reclothing such denuded land and for arresting the progress of gully formation, aforestation scheme should been taken up.

SOILS :

In the study of land use the significance of soil cannot be neglected. In fact for a detailed planning for proper land use soil survey of a region warrants a careful study. It is necessary to know the physical properties, texture, capacity of retain water and colour have much to do with the capacity and ability of the soil to produce crops as such soil survey is considered as the starting poring for all long terms planning of lands use. Mr. Chaudhary has rightly observed that the soil survey for a planning of land utilization is an inventory of certain physical condition of land of fundamental importance is determining proper land use. The importance of soil multiplies in agricultural dominated region. Even during early years cultivations used to move in search of fertile land before setting permanent in a region. Even these days availability of fertile soil becomes much more helpful to cultivators because in such soils farmers spend very little on manures and fertilizers. The importance of soil in term of cultivating is assessed in the following facts :

I. It provides physical supports to the vegetation.

II. It serves a medium for storing water and bringing water into contact with roots.

III. It supplies a small but essential percentage of the material, which is converted into food plants by means of light. As such P.H value of soil is found out. The letter P stands for power and H stands for active Hydrogen and PH means the activity of hydrogen in the soil. Each plants has an optimum PH. Range in which it produces its best some have a wide range of adoption. The Government of India is trying her best to obtain scientific data on the soils of each and every corner of the country. So that farmer can farmer can fertilizers grow suitable crops. In spite of much efforts only inadequate information are available. The detailed soil maps are not available as such the sources of information available to the author are the settlement report, district gazetteer and information gathered from farmers during field work.

Table - 2.7
Optimum pH Ranges for Different Crops

Crops	**pH Range**
Rice	5.0 - 6.5
Maize	5.9 - 7.5
Wheat	5.5 - 7.5
Mustard	5.5 - 7.0
Potato	4.2 - 7.0
Sugarcane	6.8 - 8.0

SOURCE: DISTRICT STATISTICAL OFFICE DHANBAD

During the course of field work the author visited all the selected villages of the region and information concerning texture, colour and other element of soil was gathered. Classification of soil on the basis of information gathered from the fieldwork. The soils of the study area is divided into four parts

1. Stony and Gravely soil
2. Sandy soil
3. Loamy soil
4. Clayey soil

1. STONY AND GRAVELY SOIL : These soils are found near the foot of hillock, which has a large admixture large fragment of stone, gravels pebbles etc. This type of soil may be classified as low grade soil. This type of soil is not having the capacity to retain moisture in that why it lacks vegetation; these soils are very loose so it cannot give much supports to the roots of the vegetation. This type of soil is found in the northern part of the study area and near Dhangi hills mainly.

2. SANDY SOIL : These soils are locally known as bali in the district. This type of soil is found near river and stream beds. Soils containing more than 60% of sand are classified and are easily drained as they led the water through too readily and the needs frequent watering. These soil are poor in support of plant food and require heavy manuring and in frequent doses on account of dearth of water and manure. Sandy soils are described as hungry soil. Cattle manure and composed, green manure and addition of tanks silt and clay bring about great improvement in the retentive capacity of the soil. The soils are used for growing cucurbits.

3. LOAMY SOIL : This type of soil is found near the hills and formed by rain washing from higher position and consists of detritus of decomposed rocks and vegetable matter. Soils whose compounds are between 30 to 60%

are classed as loamy soil. Agriculturally these soils are best adopted for cultivation. They are suited to every kind of crops but in the district this soils foot under Paddy, Sugarcane, Ragi, Wheat etc.

4. CLAYEY SOIL : Such soils are found near tank bed when moist they are sticky ploughing and other tillage in that condition will reduce these into of a pasty mass . When they are dry they become very hard and difficult to break. They are difficult to train, as the water cannot pass through easily on account of fineness of particles composing them. They have a high water holding capacity and a very fertile in respect of plant food content. In addition of sand, lime and coarse bulky organic manure will improve their condition. Nitrogen applied as organic matter and that in shape of ammonia free or combined as in the ammonium salt applied as manure becomes fixed in the soil i.e. that do not pass out of the soil in drainage water.

SOIL EROSION :

Soil erosion is a great problem for fertile land. Due to faulty method of cultivation due to running water and the fertile soil is eroded away. The removal of top soil either by running water of wind or other agencies in proportion more than the soil can be formed by natural process is called soil erosion. Nature generally tends to maintain equilibrium between the soil erosion and soil formation.[4] Former kind of soil is greatest wealth. The effect of soil is farfetched upon agriculture. The effects of soil erosion are common. Rill erosion is noticed along steep slope hilly area. During the rainy season almost all river of the region of full volume of water and swift current erode the bank of both sides.

The cause of soil erosion in this district may be enumerated as given below :

I. Rainfall : The area is a rain fed area; the total average rainfall is 1397 mm. The intensity of precipitation is concentrated in a short period and in a short time.

II. Running water: Due to deep slopes and high density of precipitation the run off water runs at a speed, which removes the soil.

III. Slope : The slopes of the land of the district are quite prominent.

IV. Nature of the soil: Once the runoff water makes the path. In the literate type of soil is proceeds rapidly and assumes serious proportion till gullies reaches the bed rock.

V. Ploughing : Deep ploughing is not possible due to rocky construction of the soil and as such moisture is not retained in the subsoil, but are allowed to drain. The cultivators practice ploughing along the slope. On tand land which induces and accelerates soil erosion.

VII. Lack of contour bounding which leads to a quick runoff the rainwater causing erosion.

Soil is the real wealth of the cultivation. Hence soil conservation is essential. Soil erosion is a great problem and for the well being of the people. Soil conservation is essentially needed. To conserve the soil it is necessary to conserve the farm economy. It is more important for an area having farm economy because the soil conservation is cultivation for future. When the soil lost, prosperity and culture are also lost. The soil is not dead resource. It is a living organism and not mere earth[5].

In the area under study the government is keen to solve the problem of soil erosion by tree plantation in each Anchal through the department of forest as well as through other agencies such as national social service. Scientific method of cultivation and also by tree plantation near river banks is essential in order to check erosion during the rainy season.

POPULATION :

In the study of land utilization the importance of population analysis has been considered essential by the most the scholars because all these aspects are closely related.

Since time immemorial people all over the world have tried to make use of land and keep concern with suitable land surface. The land surface of the earth has been men's home hearth and all having. Not all area have the same bearing capacity and as such land of different areas being characterized by topographic features, ecological conditions have influenced human distribution. Whipple has rightly remarks that the real wealth of any area lies not is its lands and waters, not in its forests and mines and not its flocks and herds, not in its dollars, but in its healthy and happy men, women and children.

Considering the relationship between man and land use, emphasis has been laid on detailed account of population of the district. No doubt like other geographical regions of the country this region has little authentic account early population. The reliable population data of the region are only available after 1901 although the present district came to being in the 1956. Prior to the data is part of Manbhum districts. In the area 1951 this districts has about 9.3 lakh of population. In the year 1961 the population increased to about 11.9 lakh and the year 1971 it increased to 21.2 lakh and accordingly to 1991 it increased to 26.8 lakh after division in 1995 population is lakh in the year 2001 and 26.9 lakh in the year 2011.

Table - 2.8
Variation in Population since 1901

Years	Persons	Decadal variation	Percentage of variation
1901	309930	-	-
1911	5146888	124758	31.99
1921	578512	638224	12.4
1931	670437	91925	15.89
1941	759799	89362	15.33
1951	924764	164965	21.71
1961	1182837	250073	27.91
1971	1466417	283580	23.97

1981	2115010	648593	29.28
1991	2674651	559641	26.46
2001	1949526	-	22.95
2011	2684487	734961	12.0

Source: Census handbook, 2011

As mentioned earlier the population data of the newly created districts are not available earlier than 1951. As a result growth of population of the study area has been analysed after 1951 and the population version data of the district revel the fact that the region experienced content incensing trend in the population. In between 1951 and1961 the region experienced population variation 30.68% while the rate of population in the following decades went on increasing from 1961 to 1971 the growth rate of population increased to 19.38%, in the decade 1971 it has increased to 39.18% and in the 1981-91 it has increased to 22.74%. On the hole population of region has increased by 166.5% during last 60 years (Table 2.10).

The analysis of population version of deferent Anchal also revel the fact that the growth of population is not uniform through the region. It has always fluctuated. The population of region increased by 266.83% for the last 60 year but Anchal where growth recored much more than region average are Dhanbad Anchal 318.30, Nirsa 249.97 the remaining Anchal experienced population increased less then the regional average.

The growth rate has fluctuated from decade. In the decade 1951-61 max. population variation was recorded by Nirsa (36.73) Jharia (39.21) Bhaghmara (35.73) Dhanbad (29.76) etc. The district experienced a change of 30.68% between 1971-81. Maximum variation was recorded by Dhanbad (51.65), Jharia (42.51), Nirsa (36.66), and Baghmara (42.61). Only three Anchal experienced a growth lower then the district average of 39.18%. These Anchals were Govindpur, Tundi and Topchanchi. Among these Anchal Tundi recorded the lowest growth of 17.91%. A compersion

Table - 2.9
Growth of Population of Dhanbad District 1951-2011

Block	1951	1961	1971	1981	1991	2001	2011
Dhanbad	137784	178787	223648	339158	433439	564468	1221356
Jharia	154195	214531	252338	359609	426415	475341	*Merge with Dhanbad
Govindpur	64648	78526	95987	122118	157881	201876	245697
Baliapur	42882	44324	52406	69071	84802	105613	140908
Nirsa	121897	166671	193860	264924	331133	376843	426613
Tundi	50319	59548	70540	83175	102834	124126	152262 With Purbi Tundi
Topchanchi	45744	58759	73346	94821	111024	140378	163342
Baghmara	114321	155167	179361	255919	302598	408457	334309
Dhanbad District	731790	956313	1141486	1588795	1950126	2397102	2684487

Source : census handbook, 2011

**Jharia-cum-Patherdih-cum-Sindri C.D. block- merge with Dhanbad.*

Table - 2.10
Percentage variations in the population of Anchals of the Dhanbad District since 1951-2011 Variation

Block	1951-1961	1961-1971	1971-1981	1981-1991	1991-2001	2001-11	1951-2011
Dhanbad	29.76	25.09	51.65	27.79	30.23	116.37	318.30
Jharia	39.21	17.62	42.51	18.57	11.47*Merge with Dhanbad		
Govindpur	21.47	22.23	27.22	29.28	27.86	21.70	280.05
Baliapur	3.36	18.23	31.8	22.77	24.54	33.41	228.59
Nirsa	36.73	16.31	36.66	24.99	13.80	13.20	249.97
Tundi	18.34	18.45	17.91	23.63	20.70	22.66	202.59
Topchanchi	28.41	24.82	29.28	17.08	26.43	16.35	257.07
Baghmara	35.73	15.54	42.68	18.23	34.98	18.15	192.43
Dhanbad District	30.68	19.36	39.18	22.74	22.92	11.98	266.83

Source : census handbook, 2011

**Jharia-cum-Patherdih-cum-Sindri C.D. block- merge with Dhanbad.*

between the variation of population during 1951-61 and 1971-81 reveals that Tundi has experienced the lowest growth in both decades. Among all Anchal of Danbad districts. This is attributed to the fact that the Anchal is situated in hilly area, where land is less fertile. Mining and industrial activates are less then. While Dhanbad, Nirsa, Jharia, Baghmara Anchals expedited more population by development of mining and industries.

During 1981-91 also the rate of growth of population of different Anchal has not been uniform. The district as a whole has shown a growth of 22.74% during his decade. In the analysis of the regional variation in the growth of population shows that it has been more than the district average viz Dhanbad, Jharia, Nirsa, Balipur, Baghmara. Among them Dhanbad Anchal has witnessed a growth of her higher. The figure shows that Jharia, Dhanbad, Baghmara are far above than the other Anchals (Table no. 2.10). Among the anchals experienced a low growth in the compression with district average are Tundi, Topchanchi, Govindpur and Baliapur.

A look at Anchal maps of Dhanbad shows that southern part of district experience and accelerated rate of growth, where the development of mining has been remarkable Northern part of the district is hilly has naturally not seen a growth of agriculture and economical activates. Transport facilities are less provided.

DENSITY OF POPULATION :

Due to mining and industrial development study area have high density of population with an area 2040 Sq. Km and a population of 2684487; district has a density 1316 person per Sq. Km as for 2011 census. This density is far more than the density of Jharkhand state, where in one Sq. Km only 414 people reside. Again this figure is 3 times more than the all India density. This speaks highly of the density of population of Dhanbad district.

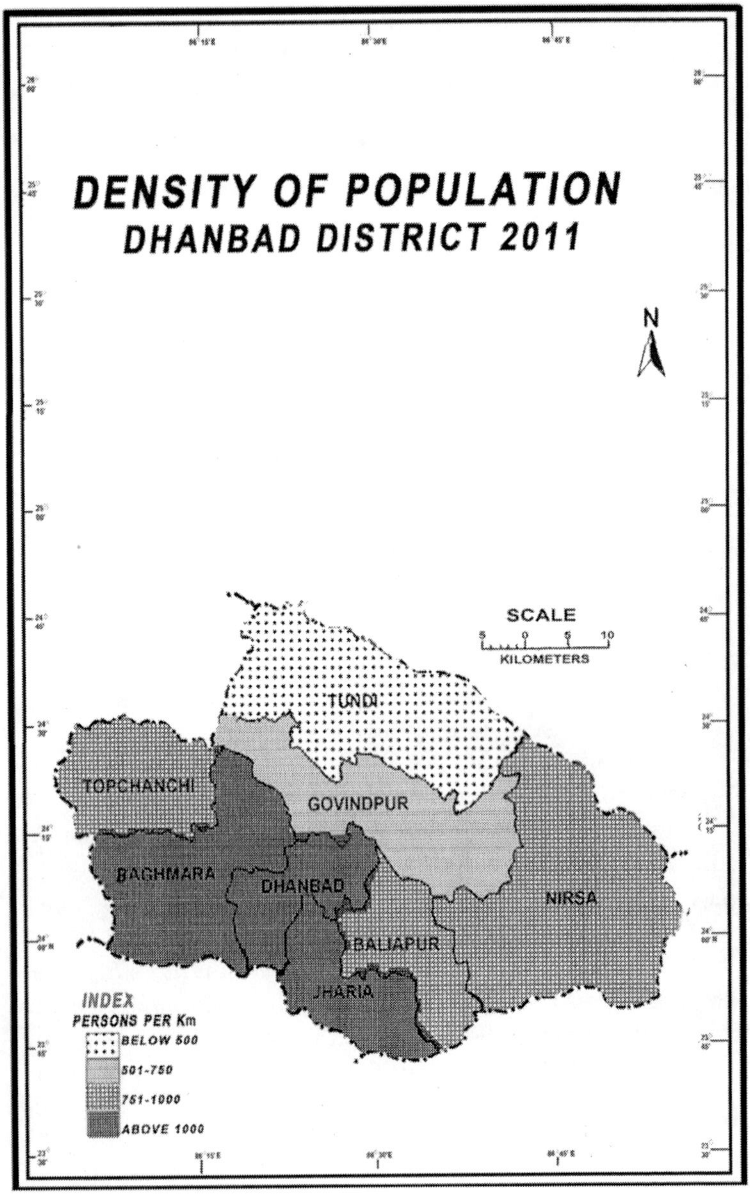

Fig. 2.6

Source : The Map Prepared by Author

Table - 2.11
Density of Population

Block	Area	1971	1981	1991	2001	2011
Dhanbad	64.20	1672	2535	3239	8792	8295
Jharia	83.03	1981	2823	3347	5724	*Merge with Dhanbad
Govindpur	329.71	287	365	472	612	745
Tundi	495.39	178	210	260	250	307
Nirsa	431.91	436	595	744	872	988
Baliapur	173.31	339	442	543	609	813
Baghmara	267.55	635	906	1072	1526	1250
Topchanchi	194.90	368	476	557	720	838
Dhanbad District	2040	550	766	940	1175	1316

Source: Census handbook, 2011

The density of population is not uniform in all parts of the district. Due to industrial factor the southern part of district had the highest density of population since 1916. In 2011 Dhanbad Anchal and Jharia Anchal had the highest density of population. Dhanbad 82.95 person per Sq. Kms had far highest density which was followed by Baghmara, Nirsa. The Anchal having lowest density of population were Govindpur, Topchanchi, Tundi and Baliapur having density of 745, 838, 307 and 813 person per Sq.km respectively (Table no. 2.11)

SPATIAL DISTRIBUTION :

As regards spatial distribution of population, it is significant to note that the population is not distributed evenly through out the district. A comparison between the figures showing the share of Anchals in area and population shows that the largest Anchal has not the largest shared of the population. Dhanbad Anchal covering 6.5% of area of the district has largest shared of population. Jharia Anchal covering 6.1% of area the district has also highest share of

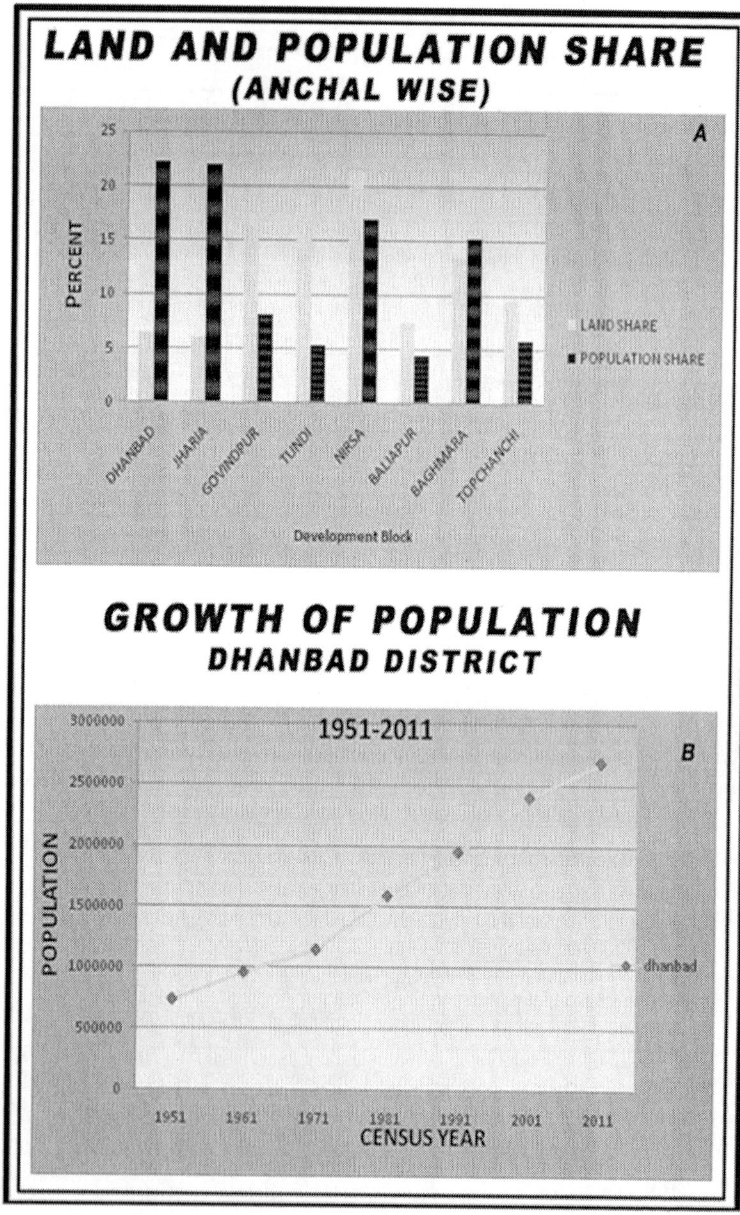

Fig. 2.7

Source : The Map Prepared by Author

population while Tundi covering 19.1% of land has only 5.3% of population. Nirsha is largest Anchal has third position in term of population share. Baliapur and Topchanchi has lowest share of land share and population share. (Table No. 2.12 & Fig No. 2.7)

Table - 2.12
Land and Population share anchalwise

Block	Land share percentage	Population share percentage
Dhanbad	6.5	22.2
Jharia	6.1	22
Govindpur	16.1	8.1
Tundi	19.1	5.3
Nirsa	21.5	17
Baliapur	7.5	4.4
Baghmara	13.6	15.2
Topchanchi	9.6	5.8
Total	**100**	**100**

Source: Census handbook, 2011

LITERACY :

From the data present in the table 2.13, it can be seen that for the district as a whole, the literacy rate of males is much higher than that of females. The literacy rate of males is as 83.81% as against 64.29% for female, the difference of almost 20% points. The C.D. blocks wise mate literacy rates shows that Dhanbad- cum- Kenduadih- cum- Jagta C.D with 86.26% literate is at the top very closely followed by Topchanchi C.D. block with 85.13% literate. The difference of literacy rate between the highest literacy rate in Dhanbad-cum-Kenduadih- cum-Jagta-cum-jorapokhar-cum- Sindri block are presumably on account of important mining and industrial town located in these blocks where adequate educational infrastructure might be available. It is also observed that some C.D. block which do not consist of any

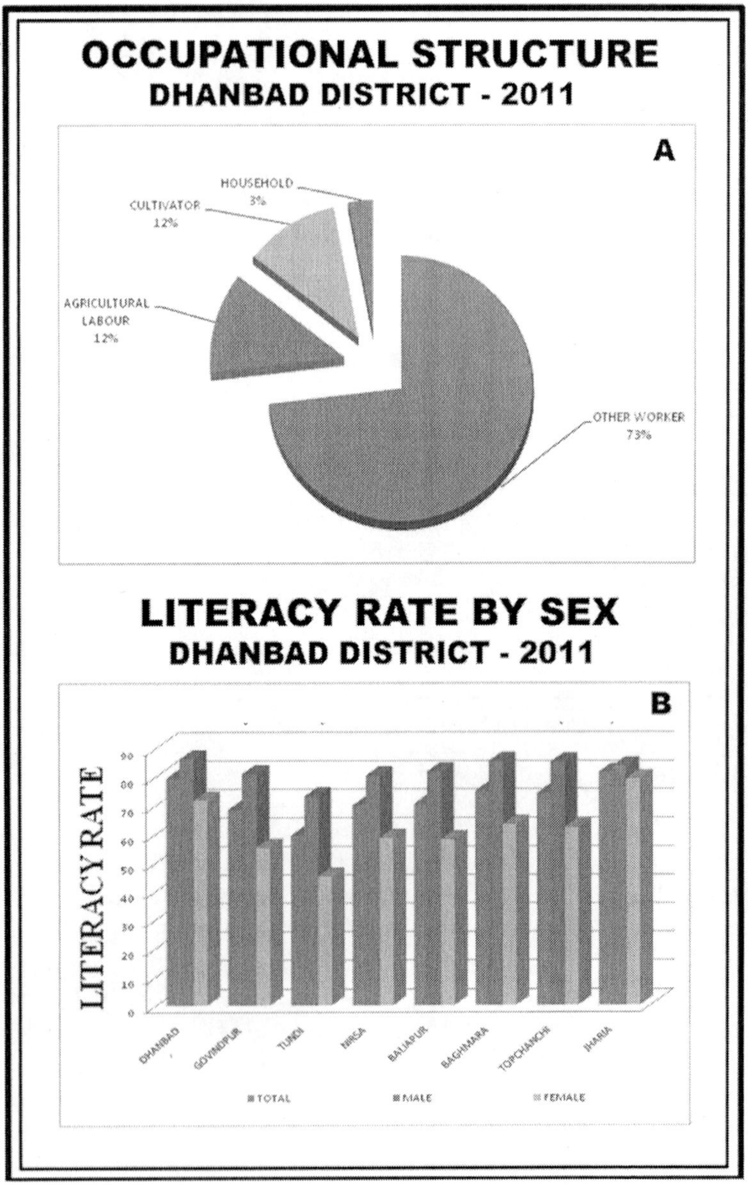

Fig. 2.8

Source : The Map Prepared by Author

urban component corroborating the fact that urban centres have better infrastructure as far as education facilities are concerned. Tundi block has registered lowest literacy rate in the district 45.5% perhaps may be due to the fact that it is deprived of proper education infrastructure, because of beging a backward area. It is also interesting to note that expects Dhanbad-cum-jorapokhar-cum-Sindri blocks, the literacy rate in all other C.D. block is lower than that district average.

As regard rural area of the district the literacy rate is observed to be 68.17 percent with male literacy rate being 80.53% and female literacy accounting for 54.9% on the other hand the literacy rate in the urban area of the district is to the extent of 78.97% the male and female literacy rate being 86.06 and 71.03 percent respectively.

Table - 2.13
Literacy Anchal wise - 2011

Block	Male Population	Female Population	% of Total	% of Male	% of Female	Total
Dhanbad	486394	359754	79.42	86.26	71.73	846148
Jharia	-	-	-	-	-	-
Govindpur	86332	54901	68.53	80.96	55.21	141233
Tundi	31415	18493	59.43	73.08	45.11	49908
Nirsa	152754	103187	69.86	80.34	58.56	255941
Baliapur	51309	33674	70.32	81.67	58.03	84983
Baghmara	128719	86169	74.92	85.46	63.26	21488
Topchanchi	61651	41650	74.1	85.13	62.18	103301
Dhanbad District	1014950	707254	74.52	83.81	64.29	1722204

Source: Census handbook, 2011

SEX RATIO :

In Census, as we know, sex ratio is defined as number of females per 1000 males. If number of females is less than 1000, it is said that sex ratio is unfavorable for females. It transpires from Table 2.15 above that the sex ratio for the rural as well as the urban areas of Dhanbad district has comparatively been much lower than the sex ratio of the state right from Census year 1901 to 2011 Census. During post Independence era the sex ratio of the state had shown downward trend. So far sex ratio in the district is concerned, the table reveals that except in the Census year 1901, when the sex ratio was 951 (somewhat not so unfavorable for females), the sex ratio since 1911 to the present Census was quite unsatisfactory and adverse for females, fluctuating between one Census year to another. However, during 2011 Census it has shown some improvement as sex ratio recorded during this period is higher vis-à-vis 2001 Census both for the state and the district, in the rural as well as urban areas, which is an encouraging trend, though sex ratio of the district is far behind the corresponding figures of the state.

The sex ratio of the state of Jharkhand in total is 949 females per 1000 males whereas it is 909 females per 1000 males in respect of Dhanbad district. Sex ratio of the state is 961 in rural areas and 910 in urban areas where as it is 932 in rural areas and 894 in urban areas in respect of Dhanbad district. It may be seen from above that in Dhanbad district there are, on an average, 909 females for every 1000 males. Among the C.D. Blocks, Purbi Tundi has recorded the highest sex ratio at the time of 2011 Census (960). It is followed by Tundi (955). It may be noted that Tundi and Purbi Tundi blocks do not contain any urban component and is totally rural in nature. It has generally been observed that the sex ratio of the urban sectors are more adverse for females as compared to the rural areas. The sex ratio of the rural areas of the district has been returned as 932 in 2011 Census as against only 894 (females per 1000 males) for the urban areas, the difference among the two being over 38 points.

Study area is an industrial region. So 49.35% people living in the urban area and 50.65 percent people living in the rural area. In compression with India the percentage of urban population very high. Urbanization has rapidly increased during last 50 years. The distribution of urban of population is very uneven. Jharia is the largest urban center having 405249. Dhanbad Anchal has 353620 urban populations. Rural

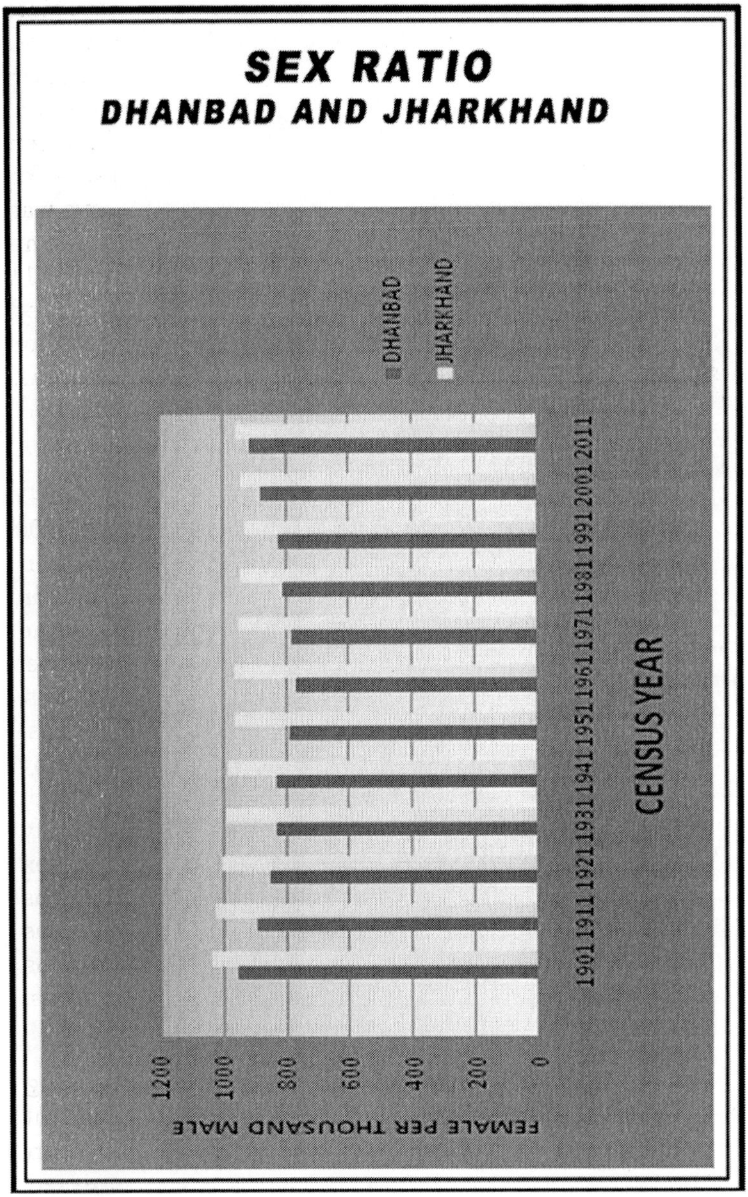

Fig. 2.9

Source : The Map Prepared by Author

population in Tundi Anchal is 100%. In this Anchal have nothing-urban population. (Table no. 2.15)

Table - 2.14
SEX RATIO OF STATE AND DISTRICT, 1901-2011

Census	Jharkhand			District Dhanbad		
Years	Total	Rural	Urban	Total	Rural	Urban
1901	1032	1034	955	951	951	0
1911	1021	1024	913	891	891	0
1921	1002	1010	809	849	857	584
1931	989	998	792	831	837	655
1941	978	990	792	827	839	666
1951	961	976	803	787	806	632
1961	960	987	772	763	817	647
1971	945	982	769	780	880	678
1981	940	977	804	807	881	740
1991	922	950	826	819	871	768
2001	941	962	870	874	909	844

Note : Sex ratio has been defined here as the number of female per 1000 males.

Source: Census handbook, 2011

Table - 2.15
SEX RATIO BY C.D. BLOCKS, 2011

Sl. No.	Name of Sub-district	Sex ratio Total	Rural	Urban
1	Tundi	955	955	0
2	Purbi Tundi	960	960	0
3	Topchanchi	922	924	919
4	Baghmara-Cum-Katras	905	913	889
5	Govindpur	930	933	908
6	Dhanbad-Cum-Kenduadih-Cum-Jagata	891	920	891
7	Baliapur	923	924	918
8	Nirsa-Cum-Chirkunda	927	940	907
	District Dhanbad	909	932	894

Source: Census handbook, 2011

Table - 2.16
RURAL URBAN POPULATION 2011

Sl. No.	Block	Population 2011 Total	Rural	Urban
1	Tundi	102022	102022	0
2	Purbi Tundi	50240	50240	0
3	Topchanchi	163342	111982	51360
4	Baghmara-Cum-Katras	334309	222657	111652
5	Govindpur	245697	220384	25313
6	Dhanbad-Cum-Kenduadih-Cum-Jagata	1221356	33421	1187935
7	Jharia-Cum-Jorapokhar-Cum-Sindri*	0	0	0
8	Baliapur	140908	121494	19414
9	Nirsa-Cum-Chirkunda	426613	261893	164720
	District Dhanbad	2684487	1124093	1560394

Source: Census handbook, 2011

OCCUPATIONAL STRUCTURE :

The relative importance of the main spheres of economic activity may be gauged from the pattern of distribution of workers (main + marginal) according to the broad fourfold classification namely, cultivators, agricultural labourers, household industry workers and other workers. The economic activity in the district is not much dependent on agriculture as a big chunk of working force are associated in non-agricultural pursuits which is supported by the fact that the workers engaged in "other works" i.e. various types of work other than agricultural activities and household industries account for as much as 76.0 percent of the total working population. Cultivators (11.50%) and agricultural labourers (12.25%) together constitute around 24 percent of the total workers of the district. The primary sector i.e., the agricultural sector has absorbed a little less than one fourth of the total workers of the district. Household industry workers are meagre 3.07 percent of the total workers.

Table - 2.17

Distribution of workers by sex in four categories of economic activity in Sub- District, 2011

Sl.	Name of Sub-District	Persons/ Males/ Females	Total Population	Total workers (Main+ Marginal workers)	Category of Workers							
					Cultivators		Agricultural Labourers		Household industry workers		Other workers	
					Number	%	Number	%	Number	%	Number	%
1	2	3	4	5	6	7	8	9	10	11	12	13
1	02579-Tundi	Persons	102022	43341	10969	25.31	18492	42.67	1839	4.24	12041	27.78
		Males	52181	26260	6246	23.79	9289	35.37	867	3.30	9858	37.54
		Females	49841	17081	4723	27.65	9203	53.88	972	5.69	2183	12.78
2	02580-Pundi-Tundi*	Persons	50240	25364	8465	33.37	10559	41.63	766	3.02	5574	21.98
		Males	25634	14099	4199	29.78	4833	34.28	474	3.36	4593	32.58
		Females	24606	11265	4266	37.87	5726	50.83	292	2.59	981	8.71
3	02581-Topchanchi	Persons	163342	54855	9381	17.10	10653	19.42	2060	3.76	32761	59.72
		Males	84974	39996	4821	12.05	5826	14.57	1184	2.96	28165	70.42
		Females	78368	14859	4560	30.69	4827	32.49	876	5.90	4596	30.93
4	02582-Baghmara-Cum-Katras	Persons	334309	94817	7276	7.67	6594	6.95	1953	2.06	78994	83.31
		Males	175479	78216	4747	6.07	3853	4.93	1283	1.64	68333	87.36
		Females	158830	16601	2529	15.23	2741	16.51	670	4.04	10661	64.22
5	02583-Govindpur	Persons	245697	99119	24038	24.25	18123	18.28	4182	4.22	52776	53.25
		Males	127285	64490	10881	16.87	7722	11.97	1957	3.03	43930	68.12
		Females	118412	34629	13157	37.99	10401	30.04	2225	6.43	8846	25.55

6	02584-Dhanbad-Cum-Kenduadih-Cum-Jagnta	Persons	1221356	335629	3372	1.00	6238	1.86	9108	2.71	3E+05	94.42
		Males	645762	290573	2443	0.84	4836	1.66	6793	2.34	3E+05	95.16
		Females	575594	45056	929	2.06	1402	3.11	2315	5.14	40410	89.69
7	02586-Baliapur	Persons	140908	47231	10090	21.36	6294	13.38	2052	4.34	28795	60.97
		Males	73278	35206	6048	17.18	3838	10.90	1147	3.26	24173	68.66
		Females	67630	12025	4042	33.61	2456	20.42	905	7.53	4622	38.44
8	02587-Nirsa-Cum-Chirkunda	Persons	426613	144148	23527	16.32	26535	18.41	3929	2.73	90157	62.64
		Males	221363	110011	13940	12.67	15482	14.07	2662	2.42	77927	70.84
		Females	205250	34137	9587	28.08	11053	32.38	1267	3.71	12230	35.83
	District : Dhanbad (354)	Persons	2684487	844504	97118	11.50	103488	12.25	25889	2.07	6E+05	73.18
		Males	1405956	658851	53325	8.09	55679	8.45	16367	2.48	5E+05	80.97
		Females	1278531	185653	43793	23.59	47809	25.75	9522	5.13	84529	45.53

The situation is, however, quite different among various C.D. Blocks Dhanbad-cum-Kenduadih-cum-Jagta is the one block in the district where the proportion of workers engaged in "other works" is quite high, being around 95 percent. Baghmara-cum-Katras block has also registered quite high percentage of 'other workers' (83.31%). This may presumably be on account of large number of people engaged in coal fields and other technical jobs in various industrial and allied natures of establishments.

It can further be seen that household industry workers constitute comparatively smaller proportion (3.07%) in the district of Dhanbad.

TRANSPORT :

Transport influences the economy as also land use of a region. Its role is the economic development is as important as that of the blood circulation in the body. The distinct is well developed from the point of communication. The grand truck road runs through the entire length of the district form west to east and divided it into two parts i.e. northern parts and southern parts. The area of the south is well served by roads as well as railways. Whereas the northern portion, which consists of hill is deprived of adequate communication facility. The following are the roads maintained by P.W.D department in the district.

1. Grand Truck Road (N.H) - 72 km
2. Govindpur-Jamshedpur road - 40 km
3. Govindpur-Baliapur-Sindri road - 24 km
4. Dhanbad-Patherdih-Sindri road - 26 km
5. Rajganj-Katras-Jamdiha road - 20 km
6. Giridih-Tundi road - 28 km
7. Jharia-Baliapur road - 28 km
8. Above mention all roads are metalled. Beside the P.W.D.

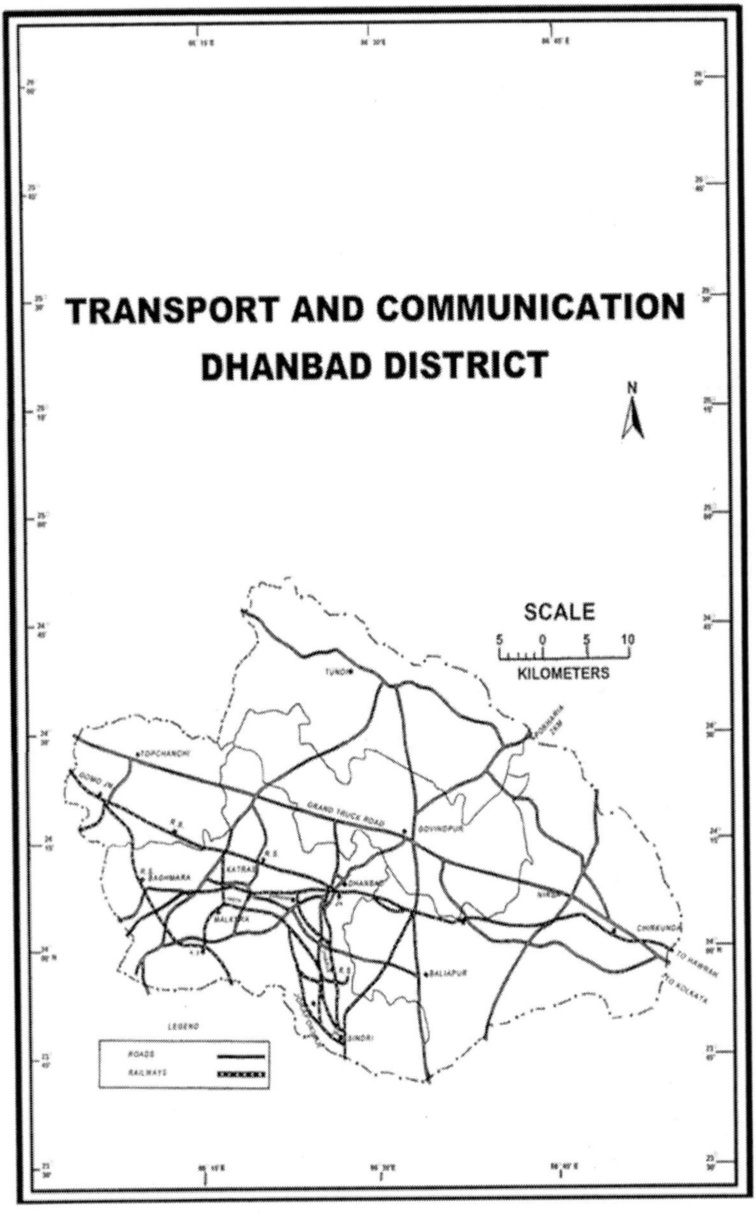

Fig. 2.10

Source : The Map Prepared by Author

few roads in the district also mentioned by zila parisad, the Dhanbad municipality and the forest department.

9. The district is well served by railways. There is a network of railway lines running through colliery area. Important among there being.

10. (i) Kumardhuby Gomoh section- it is a part of the Grand Chord section under the eastern railways. It connected Dhanbad with the grand chord main line running from Howrah to Delhi via Asansol, Gomoh, Gaya and Mughal Sarai.

    (ii) Pradhankhanta- Patherdih section

    (iii) Dhanbad- Purulia section

    (iv) Dhanbad-Patherdih section

    (v) Sindri-Gomoh section

Besides the above there are other branch lines for carrying coal from one place of the district to the another. The whole of the middle portion of the district is thus virtually infested with railways.

No river in the district is navigable as most of them are hill streams which remain dry except during rain.

The Dhanbad is not on air route but there is a small landing ground about 4 km north of Dhanbad district. There are several aerial ropeway used for transportation sand to different collieries.

REFERENCE :

1. Dhanbad district Gazatteer 1964, p.2
2. Dhanbad district census handbook 2011
3. Loo is a Hot and dry wind which is very dangerous to Human being as well as animals.
4. Sharpe, C.F. (1938), "What is soil erosion" U.S depts. of agriculture, misc., pub. 286, p. 1.
5. M. Prasad, "The Changing Pattern of Land Use in Bodh Gaya Block" Unpublished Ph.D. Thesis, p. 69

3
Chapter
Measurement and Classification of General Land Use Pattern

The purpose of land use classification is to maximize the productivity and to conserve the land for prosperity. A classification of land requires considerable time and expense. There for, there must be a justification to classify any land into different categories. Land classification not an end in its self but a means of obtaining better land use. The primary aims of the general pattern of land use study is to portray the plan and distribution of land to various uses ranging from field to factory, forest to foundry, pasture to pond and sandy stretch to settlement.

The pattern of land use is influenced by two main factor : (i) Physical and (ii) Human. Physical factors include topography, climates and soils, which set the broad limits upon. The capabilities of the land and the human factors are the density, occupation of people the technological level and socio economic institutions which determine the extent which the physical capacity of land are utilized. In the previous chapter above are described in detail. Now in this chapter describing about different categories of land to know their utilization in the study area.

CLASIFICATION OF LAND :

Land classification means dividing the land into different categories according to a single factor or a set of factors. Therefore, land utilization may be of many types depending upon the factors taken for classification. The land classification relates to climatic factors soil characteristics, slope of the land, degree of erosion, water supply, drainage and similar environmental condition. The land use capabilities classification portrays the physical capabilities of land to produce over a long period of time under stated condition of uses and which can provide land operation with a basis for actual practices on[1] scientific unit of land.

Any classification of land is any part of the world is guided primarily by the classification made by stamp, who is his pioneer work.[2] "The land of Britain: - its uses and misuses" has classified land into six major categories. This is also called the British land use classification the major groups are (i) Forest and woodland (ii) Arable land (iii) Meadow land and permanent grass (iv) Heath and moor land (v) Garden, orchard nurseries and (vi) Unproductive land, such as building, mines waste land etc.

In the international land use classification nine majors land use classes are recognized. (i) Settlement and associated non-agricultural land (ii) Horticulture (iii) Trees and permanent crop (iv) crop land (v) Improved permanent pasture (vi) Improved grazing land (vii) Woodlands (viii) Swamps and marshes (ix) unproductive land.

In the United States of America the widely used scheme of land use classification are based on land capability classes. Three classes are generally determined which are based on land the local relief, climates, soil and vegetation. The land is then classified in three categories according to anticipated results of different type of use : (i) Land capable of carrying various crops or grass in rotation or permanent (ii) Land suitable for the poorer grades of permanent pasture and (iii) Land unsuitable for economic use as arable, grazing or a forestation. The soil conservation service and the agricultural land use planning activities of the U.S.A department of agriculture had classified the land into four capabilities classes. The national resource planning board of land classification in the U.S.A classified the land into five categories :-

(i) Land classification in term of in the rent characteristics (ii) Land classification in terms of percent use (iii) Land classification in term of use capabilities (iv) Land classification in term of recommended use and (v) Land classification in term of programme implementation.

In the U.S.S.R. the following types of land are distinguished in the general land use map. (i) The land belonging to the urban and rural localities to the industrial and transport departments. (ii) Arable land (iii) Perennial plantations (iv) natural grasslands (v) Forests (vi) Bogs and peat bogs and (vii) Unsuitable lands. A.K.Phil brick has classified the land of china in to the following categories : - (i) Agricultural land 60% or more under cultivation (ii) Agricultural land at least 20% under cultivation (iii) Advancing agriculture and a forestation (iv) Upland with some cultivation and patches of forests (v) Natural forests (vi) Steppe and desert grazing lands, some cultivation and patches of forest and (vii) Major drought area. Generally the land utilization of Japan is classified in to rice land, high land with dry agricultural, forested land and uncultivated land. Land use survey in Japan has been conducted since 1953 and up to the present a large number of sheets have been published. On the basis of theses land use map 11 (eleven) main categories of land have been recognized which have further been subdivided into many micro subdivision. These categories are :- (i) Paddy area (ii) Upland field (iii) Three crops (iv) Forest area (v) Grass land (vi) Others indicated by salt bed Barren land, fish culture areas and boundaries of different fields. (vii) City and town (viii) Transportation (ix) Land improvement and land conservation facilities (x) Special facilities such as power plants, mine, transformer substation and (xi) Land futures such as a swampy lands and peat bog.

In India various schemes have been worked out to classify the land into different uses. The national atlas organization in 1957 classifieds the land into nine categorize :- (i) Forests (ii) Scrub (iii) Arable land (iv) Arable land with trees (v) Plantation (vi) Pasture (vii) Waste land (viii) Alpine grass and scrub (ix) Glaciated region. In 1968 the diagnostic survey of the Damodar valley region has classified land into 10 major categories :- (i) Field crops (ii) Orchard (iii) Dense forest (iv) Light forests (v) Non agricultural use (vi) Unproductive land (vii) water bodies (viii) Culturable waste (ix) Village and (x) Cities and towns.

Before 1951 the land use statistics collected by the ministry of agriculture, government of India where arranged according to the[7] following classification. (i) Total geographical area (ii) Area under forest (iii) Not available for cultivation (iv) Current fallow (v) Other uncultivated land and (vi) Net area sown.[8] In 1954 E. Ahmad classified the rural land use into the following categories with detailed coloured technique for land use mapping (i) Triple cropped land (ii) Doubled cropped land (iii) Single cropped land (iv) Fallow land (v) Land under grove and orchard (vi) Land under scrub and grasses (vii) Land permanently under water (viii) Built up land (ix) Land near communication (x) Barren land and (xi) Forest. While classifying the land use of canal irrigated area in Patna district[9] sinha classified the land into. following categorizes (i) Single cropped land (ii) Double cropped land (iii) Triple cropped land (iv) Current follow (v) Culturable waste (vi) Waste land (vii) Forest and woodland (viii) Village grove and orchard (ix) Bushes and scarabs (x) Grasses (xi) Built of area and (xii) Water bodies.

CLASSIFICATION OF LAND DHANBAD :

The classification of land for the country as a whole is not applicable to the area under study. Study area is a plateau zone. The land utilization is determined by nature of land having marked local version in slope. Some parts of the district contain higher ground some contain lower ground, while some are fertile and others are unfertile. In this way different physical sets of land present diverse type of land use.

The general land use of Dhanbad district has been categorized in to the following nine type (i) Net area sown (ii) Land put to non agricultural use (iii) Current fallow (iv) Forest (v) Barer and uncultivated land (vi) Other fallow (vii)

Land under mixed trees and groves (viii) Cultivable waste land (ix) Permanent pastured and other grazing lands. The detailed land use category areas covered by each of the uses have been shown in (Fig. No. 3.1) & table 3.1

Table - 3.1
LAND USE PATTERN OF DHANBAD IN 2014-15

	Land use	Area in acres	% of total area of the district	% of total area of the state
I.	Net area sown	103944.38	20.62	22.7
II.	Land put to non-agricultural uses	119368.74	23.68	9.9
III.	Current fallow	67145.45	13.32	12.1
IV.	Other fallow	51568.91	10.23	12.9
V.	Forest	46780.01	9.28	29.2
VI.	Farm pasture and other Graze	1411.46	0.28	1.3
VII.	Land under trees gardens	5293.98	1.05	1.2
VIII.	Cultivable waste	28128.49	5.58	3.5
IX.	Barren and cultivable land	80453.55	15.96	7.2
X.	Land sown more than once	1032.01	00.99	2.6
	Total	**504094.97**	**100%**	-

Source: Dhanbad statistical office

Broadly we can bring all the nine different uses of land into major classes' viz. Arable ad non-arable. The net area sown current and other fallows, Groves and orchard conveniently are put under arable group. Other groups may be combined into non-arable land. Taking Dhanbad district as a whole into consideration 45.22% of its total area can be said to be arable while 54.78% is not available for agricultural purpose but out of the total percentage available land only

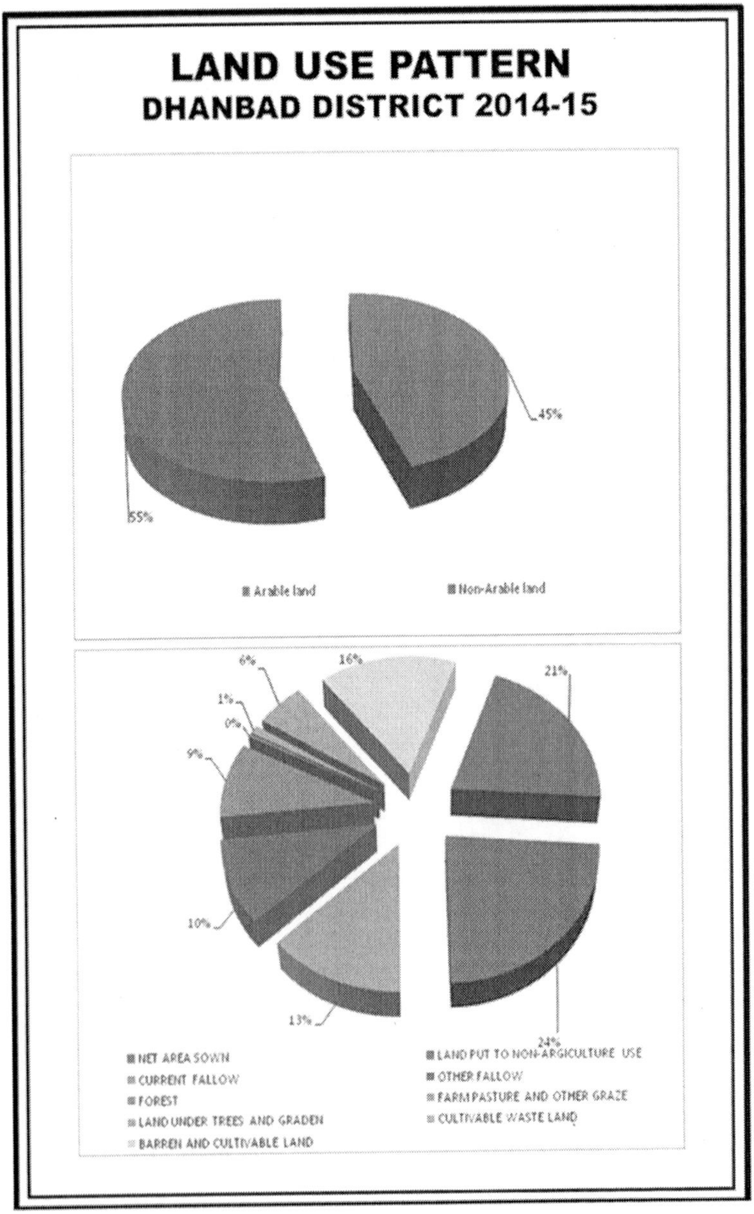

Fig. 3.1

Source : The Map Prepared by Author

20.62% was actually sown with crops. This is known as "Net Sown Area" in the government's records. It presents the geographical extent of sown or cultivated land during a particular year. "Current Fallow" is also a part of cultivated land during from the net sown in sense that at the time reporting the use covered by this had no crops in the land. Although it was cultivated only during the last agriculture year. Similarly other fallow is that type of arable land which is temporarily out of cultivation due to practice of crop rotation or some other reason. But this type of land is never left as fallow for more than five years. Thus it is justified to call it Arable land. Groves and orchard are set to be sown land under perennial and plantation crop. Because they are considered to be cash crops. Many open orchards are regularly ploughed and sown with varieties of crops.

Non-arable group of land are those, which cannot be brought under cultivation without bringing radical changes in the environment. "Cultivable waste" is virtually a wasteland owing to a number of severe limitations, which stands in its way of cultivation. This limitation may be excessive sandiness, water logging, weed jungle etc., which are necessary to be removed to bring such land under cultivation, "Barren land" includes different types of lands such as sandy lands, lands under river bed, perennial ponds and marshes which cannot be brought under cultivation. "Land put to non agricultural use" has been put to non arable uses by men himself. Among the non agricultural uses of land are those occupied by building roads buildings railways, air strips, artificial tanks and canals etc. A most remarkable constituent of Non-arable land is the area occupied by forest.

A glance at the figure of land use in Dhanbad shows that only 1/5th of its land is under net area sown. This shows less importance of agriculture in the economy of the district. This is because the whole region is plateaus and hilly, where soil is less fertile. Irrigation facilities are not available. Agriculture is provided on Bahal and Baid land mainly. Next comes land put non-agricultural use which occupies

About 23.68% of total area of the district. This shows the development of the district at this category of land. As this category of land includes building, roads railways and other man made uses. Current fallow constitutes 13.32% of the total area. This is very high figure and shows that a large area of land has been left vacant during the year of reporting. Land under other fallow constitutes 10.23% of the total area. An area under this category of land is undulating tracks and a rock slope which has compelled people to leave agriculture land for 2 to 3 years. Cultivable waste and barren and uncultivated land covers more that 21% of the study area this category of land includes useless or unproductive and unfit for cultivation. Forest and permanent pasture covered 9.3% and 0.28% of the total land respectively (As shown in the figure No. 3.1).

I. NET SOWN AREA :

Net area sown denotes the geographical extent of the cultivated or the sown land during a particular year. It measures more than one lakh acres (103944.38 acres) of land accounting for 20.62% of the total land of the district. This figure is very low where compared with the average figure of the state of Jharkhand, which is 22.7%. The low ratio of land under plough is due to unfertile nature of land. The pressure of population is in non-agricultural activities. The ratio of net sown area to the total area of the district varies considerably from region to region. Depending mainly upon the surface conditions including the fertility of the soil, availability of water for irrigation and development. Among the eight Anchal or administrative region of Dhanbad four Anchal have more net sown area than the average of the district.The Anchal having more than district average figure of net area sown area Tundi, Baliapur, Govindpur and Topchanchi. Tundi has 30% of the total area devoted to crops. Jharia, Dhanbad, Baghmara and Nirsa are the Anchal where the figures for the net area sown are lower than that of the district average. This block figure can be attributed a number of factor. Most responsible factor is mining. Jharia coalfield

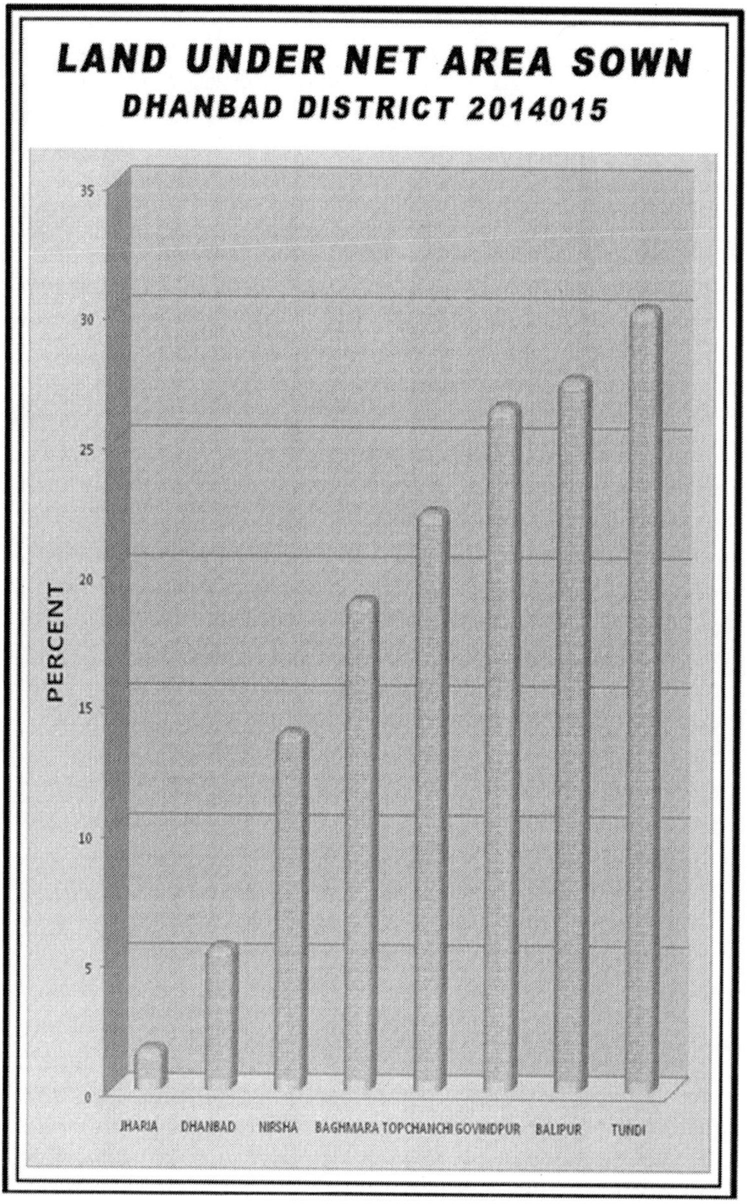

Fig. 3.2

Source : The Map Prepared by Author

belt having one of the best qualities of Bituminous coal is one of the important coal producing belts in the world. Bharat cocking coal Ltd. And Eastern coal Field Ltd. Under coal India Ltd. Has numbers of collieries. Coal mining urbanized the area rapidly. Jharia is most famous town in the district where only 1.64% of land is under net sown area. Dhanbad Anchal is the most urbanized part of the district, where Dhanbad is also district head quarter. Dhanbad town has a population of about 3 lakh. The rapid urbanization is followed by construction of buildings, roads, railways, small scale factories and educational and administrative institution has left very small acreage of land under net area sown. Urbanization has also lowered the proportion of net sown area in Nirsa and Baghmara Anchal. Where Nirsa, Chirkunda, Katras, Baghmara are important towns. All towns are centre of coal mining where most of the populations are important towns. All towns are centre of çoal mining where most of the populations are in the mining activities, commerce and trade. All towns of coalfield attract a large number of people from different part of Bihar and neighbouring state comparison of the net sown area in 1980-81 and 2014-15 shown the impact of urbanization and opencast mining on land use Dhanbad, Jharia, Nirsa and Topchanchi Anchal sown area has decreased due to construction works following urbanization on the other hand net area sown has increased in Baghmara, Tundi, Baliapur and Govindpur which shows very little urbanization. Improved irrigation facility coupled with the supply of manures and seeds and mechanization of agriculture has been responsible for an increase in the net sown area. (Table 3.2 & Fig. No. 3.2)

II. LAND PUT TO NON-AGRICULTURAL USES :

This broad category comprises of a number of different type of land which is not available for cultivation under the existing circumstances. This type of represents the land occupied by buildings, roads, railway tracks, factories, mining, water bodies etc and other land put to use other than agriculture. These lands cover 119368.74 acres and accounts for 23.68% of the total land district this use of land is first ranks in the region. When compared to non-agricultural land of Jharkhand it is very high figure as only 9.9 of the total area of the state is under this use. High proportion of non-agricultural land is due to the rapid growth of population, which requires more land for residential house, commercial establishments, educational and other institutions, industries, roads, mining, railways etc. The land under this category is increasing fast and is bound to increase in future also with the development of science and technology. This land is however considered detrimental to balanced rural agrarian economy because the productive land is usurped by unproductive uses.

Non-agricultural land is the index of development of an area when an area is developed in the construction of building and transportation facilities. This encroaches upon agricultural land. This fact is also apparent in the distributional pattern of this land in Dhanbad district. There are three Anchal Jharia, Dhanbad and Nirsa where the percentage of land under this use is higher than that of with the district average because these Anchal are relatively more developed. Jharia is main business centre of the district and having more collieries and has very high percentage of land (84.40%) under non-agricultural use. Dhanbad having the district headquarter has 33.88% of land under non?agricultural use. Apart from large residential areas, commercial establishments small scale factories, educational institution, a large part of area is covered by road transportation as Dhanbad is the node of road transport communication. Nirsa is also mining centre, considerable acreage of land is also put residential house for thick population government offices and quarter of employees and commercial establishments. This is why 23.94% of the area of Nirsa Anchal is under non-agricultural use. Impact of urbanization has resulted in construction of building road and other settlements. The population of Baghmara and Topchanchi Anchal is also high leading to an increase in the proportion of non-agricultural use. Govindpur Anchal has

21.48% of its land under non-agricultural use. It is a center of road communication where from roads go to Kolkata, Delhi and Giridih. It is also an important business centre more than 50 coke plants and other industries are situated on the both side of GT Road (Grand Trunk Road, National Highway-2). All these have combine to increase the land under non-agricultural use. The Anchal having the lowest percentage of land under non-agricultural use is Tundi. Where only 10.61% of land is under this use. It is an undeveloped part of the district where construction has been at very low level. There is no industry or any urban centre in this Anchal. Baliapur Anchal is also less developed where due to large portion land under hills construction works has been very low due to low fertility of soil the development of agriculture has been very slow leading to the sparse population. Due to this only 20.78% of the area of this Anchal is under the non-agricultural use.

A comparison between the non-agricultural land 1980-81 and 2014-15 (Table No. 3.3) show that the proportion of non-agricultural land has increased in Jharia, Dhanbad, Govindpur, Topchanchi and Baghmara Anchal where the construction was has been undertaken and completed during this period. In Tundi and Nirsa the proportion of land under this use has decreased because assured irrigation has intensified agricultural activities and agricultural use of land.

III. CURRENT FALLOW :

The definition of the term "current fallow" vary greatly in different of part of the country. "In Punjab for instant land is classified as current fallow if it has been left uncultivated for less than two years. In Maharashtra on the other hand, land continuous to be classified as current fallow if it has continued uncultivated for less than 10 years. But in Jharkhand "current fallow" is applied to all such land which was not under crops at the time of reporting but which had been sown in the recent past. Thus current fallows are a part and parcel of the arable land. The need for leaving the land fallow seems to arise when the soil is less fertile and soil

LAND PUT TO NON-AGRICULTURAL USE

DHANBAD DISTRICT 2014-15

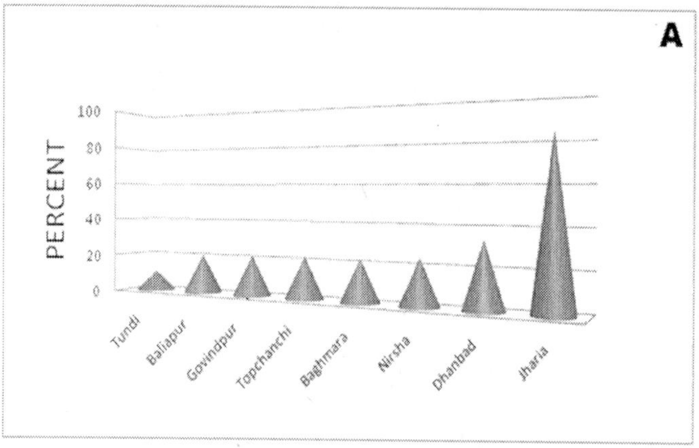

LAND UNDER CURRENT FALLOW & OTHER FALLOW

DHANBAD DISTRICT 2014-15

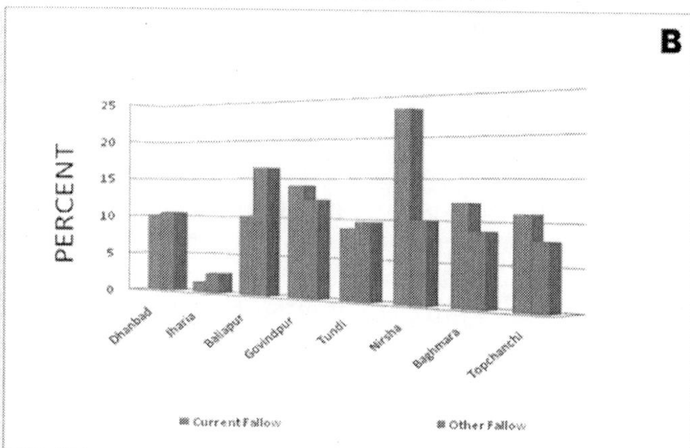

Fig. 3.3

Source : The Map Prepared by Author

exhausting crops are grown over it without much use of fertilizers.

In Dhanbad district 67145.45 acres of land are fallow currently. This works out to be about 13.4% of the total land of the district. The figure is very high as compared to that of Jharkhand state where more than 12.1% of the total land is under current fallow.

When we glance on Anchal wise distribution of current fallow (Fig. No. 3.3) It shows some remarkable features. As regards the area under current fallow the diversity among different Anchal is very large. These Anchal are Nirsa, Govindpur, Baghmara, Topchanchi have more than 10% of their land under current fallow while other Anchal account for only less than 10% of these land under this use. Lacks of irrigation of facilities and hilly tract have compiled people to leave the agriculture land fallow almost during alternate years. Due to lack of rain fall in monsoon period is main region for leave fallow and Nirsa Anchal received less rain fall in 2014-15 so it has 23.22% of land under current fallow. Tundi Anchal is mostly hilly where is found dense forest, it received more rainfall then other Anchal so only has 8.97% land under this type.

IV. OTHER FALLOW :

The land under this class is actually arable but owing to the inherent infertility of the soil and other limited factors it cannot be cultivated continuously, some short of rotation, not actually of a crops but cropping and fallowing have to be practiced for the better conservation of soil. In this practice a cropping year is followed by 2 to 5 subsequent following years. Thus such land become temporally out of cultivation for a period of not less than one year and not more than 5 years. These lands can be made profitable by using manures and irrigational facilities.

The other fallow land total 51568.91 acres representing 10.23 percent of total area of the district and 12.9 percent of the total area of the state. The high acreage of land under in

Dhanbad district is due to the fact that the area is less productive and infertile. Where undulating tracts and rocky slopes have complied people to leave agricultural land for 2 to 3 years. Lack of rainfall and irrigation facilities are main reason for having higher percentage of land under this class. Distribution of industrial ash is also factors for infertility of land. The most notable Anchal having higher percentage of land under this class is Baliapur where 16.35 percent of land is under other fallow. This is also 1.5 times more than the average percentage of land under this class in the district. Govindpur, Dhanbad, Nirsa Anchal have also cover more than 10 percentage of land under this class. Baghmara, Topchanchi, Tundi have also 8.94 percent, 7.96 percent and 9.75 percent of land under this class respectively. Jharia Anchal has 2.35 percent of land where net sown area is also a little. The presence of hilly tract and Barakar River is responsible for such category of land in Tundi Anchal and Baliapur Anchal is generally infertile sandy banks of river Damodar.

V. FOREST :

About 46780.01 acres of land of Dhanbad is covered with forest. It works out to be 9.28% of the area of the district. As compared to all Jharkhand figure of 29.09 percent it reveals very high but forest is found mostly in Tundi Anchal. Previously Dhanbad district might have been covered with forest but the growth population laid to the gradual elimination of the forest covered. The demand of timber in coal mines and fire woods caused immense loss to it. Now forests abound mostly on hill top, on deeper soils and on moderate slope. Only Jharia Anchal has no any amount of land under forest cover.

Tundi account for 27.21% of this area under forest while 15.40% of its area of Topchanchi is under forest. The forest of Tundi and Topchanchi is protected forest inter spared with trees of Sal, Mahua, Bamboo, Kenda, Gamhar, Aksia etc variety. Baghmara Anchal account for 7.38% of its area includes a scrub jungal interspersed with small trees of Palash

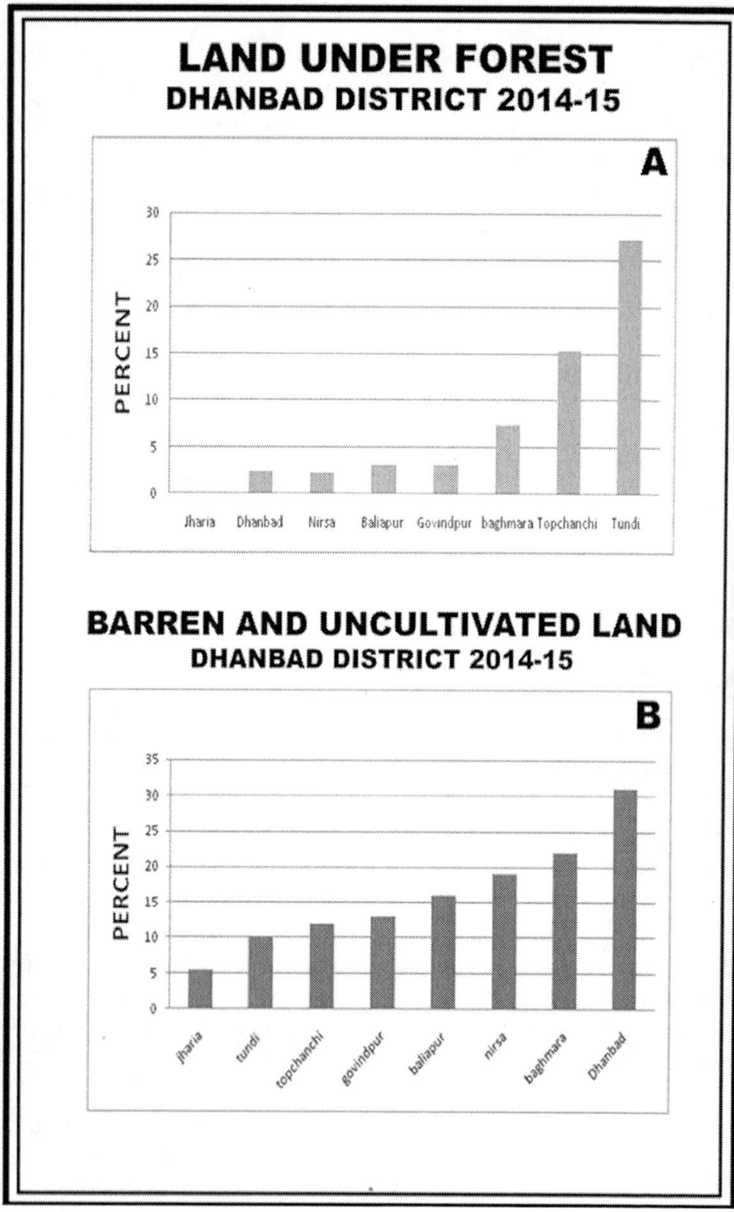

Fig. 3.4

Source : The Map Prepared by Author

and other hard woods type of trees, Thorn scrub viz putush found everywhere. Dhanbad, Baliapur, Govindpur, Nirsa have less than 5% area under forest.

VI. PERMANENT PASTURE AND OTHER GRAZING LAND :

These cover all grazing lands whether they are permanent pasture or not. Permanent pasture lands for grazing cattle had been the characteristics of land use in the past. Owing to an increase in population requiring more land for cultivation, the pasture land all over the Dhanbad has been usurped by the agriculture. Cattle grazing are done only seasonal fallow land. In Dhanbad district 1411.46 acres of land is under permanent pasture and other grazing land. This shows that the percentage of land under this category is negligible. This low percentage is mainly due to the reclamation of such area for cultivation. With the avaibility of such of irrigation water there is bound to be a rise to the growth of grass on land which is at present lying as current waste for want of moisture.

VII. LAND UNDER MISCELLANEOUS TREES AND GROOVES :

It includes all cultivable land which is not including with net sown but it is put to some agricultural uses. This category of land in arable which is not only a source of and annual plantation crops viz Mango, Guava, Mahua and Jack fruits etc. These crops contribute fairly to the total income of some farmers. Generally a large number of Mango orchards are considered to be exceptional fertile on account of high humus content supplied by the decomposition of dry leaves of the tree.

As regards Dhanbad district about 5293.98 acres of land are under miscellaneous trees and groves which account for 1.05 percent of the total area of the district. This is low percentage as compared to that of Jharkhand state when 1.2 percent of the land is under this category of use. The low acreage of land Dhanbad is district evident from its distribution pattern. Five Anchal of district are almost devoid of orchards and groves where the percentage of land under this use is negligible. Only three Anchal Govindpur, Topchanchi, Nirsa have considerable trees and groves. 2.89 percent of land of Govindpur, 1.59 percent of Topchanchi Anchal and 1.14 percent Nirsa Anchal is under groves and orchard all these Anchal account for higher than average percent of land under this use. Only .07 percent of Baliapur is however, under this use. The main trees are those of Mango, Guava, Jamun, Jake fruit etc. These are very valuable fruit of the district and supplement the otherwise weak economy of the farmers.

Due to rapid growth of population and growing needs of food product area under orchard and groves are decreasing day by day. An orchard generally provides fruits only once in a year but several years without any invest of money. Fig. No. 3.1 shows that in almost all anchals are under orchards and groves has decreased considerably.

VIII. CULTIVABLE WASTELAND :

Cultivable or Culturable waste lands are definitely cultivable but are at present lying as waste on account of number of limitation. The limitation varies from one area to another. They can be enumerated under the following heads; encroachment by wild weeds, gully erosion, poor drainage, scarcity of water, distance from settlement sites etc. As far cultural waste land of Dhanbad district it is found in the area where the land has been adversely affected by erosion and scarcity of water. Due to these negative factors this category of land has become uneconomic to cultivate. These lands can be brought under cultivation if they are protected from erosion and water logging. The cultivable waste land in Dhanbad is 28128.49 acres only about 5.58% of the total area of the district. In Jharkhand state the percentage of land under cultivable waste is 3.5%. Dhanbad, Govindpur, Jharia, Topchanchi, Nirsa, Baliapur, and Baghmara have 5.71%, 6.10%, 4.18%, 6.32%, 7.07%, 5.32%, and 7.40% of land under cultivable waste.

The existence of cultivable waste lands in the region with agrarian rural economy present much scope for extension of cultivable particularly where agricultural land is scarce. In Dhanbad district also the area under cultivable waste has decreased giving rise to extension of cultivated area during 1980-98. A glance at Fig. No. 3.1 shows that the percent of the land under cultivable waste has decreased considerable during this period in Tundi, Baghmara, Nirsa and Baliapur Anchal.

IX. BARREN AND UNCULTIVABLE LAND :

This category of land includes all such land which are practically use less or unproductive and unfit for cultivation. These lands are either rugged or sandy wastes or alkaline tracts, which cannot be brought under cultivation except at a high cost. Plant growth is possible neither of them and there are, therefore, unsuitable for cultivation or pasture. The alkaline tract can be brought under cultivation if there inundated by at so that the salt continent might dissolve and downwards.

More than 80,000 acres of land of Dhanbad which is about 15.96% of the area of the district, is barren and uncultivable. The percent of this category of land is higher in Dhanbad district where we see that only 7.02% of the total area of Jharkhand is barren. The high percent of barren and uncultivable land in Dhanbad district is due to the plateaus region and river beds throughout the district.

Dhanbad Anchal is far the most significant Anchal where about 31% of land is barren and uncultivable. It is due to fact that a large tract of the Anchal is under rugged land which cannot be cultivated due to the infertility. A part from this considerable area is under riverbeds which also cannot be brought under cultivation. Baghmara stand second with about 22% of its land under this category. This also attributed to the presence of rugged land. Topchanchi, Nirsa, Baliapur, Govindpur are other Anchal where 12.15%, 18.92%, 15.94%, 13.10% of land is barren respectively. These barren lands are

Table - 3.2

Percentage of General Land use pattern (block wise) of the Dhanbad district 2014-15

Block	Net area sown	Uncultivable Barren Land	Forest	Land put to Non-agriculture	Permanent pasture	Trees and Grooves	Cultivable waste land	Current fallow	Other fallow
Dhanbad	5.54	30.88	2.41	33.88	0.15	0.78	5.71	10.15	10.5
Jharia	1.64	5.46	-	64.4	0.11	0.31	4.18	1.37	2.53
Baliapur	27.5	25.94	3.14	20.7	0.8	0.07	5.37	10.21	16.35
Tundi	30.24	9.96	27.21	10.61	0.18	0.56	2.62	8.95	9.75
Nirsa	13.87	18.92	2.3	23.94	0.22	1.14	6.32	23.22	10.07
Baghmara	18.93	22.43	7.38	22.49	0.11	0.15	7.4	12.4	8.94
Topchanchi	22.35	12.15	15.4	22.23	0.029	1.59	7.07	10.96	7.96
Govindpur	26.45	13.1	3.1	21.48	0.4	2.89	6.1	14.11	12.37

Source : Calculated by author.

Table - 3.3
Percentage of General Land use pattern (block wise) of the Dhanbad district 1980-81

Block	Net area sown	Uncultivable Barren Land	Forest	Land put to Non-agriculture	Permanent pasture	Trees and Grooves	Cultivable waste land	Current fallow	Other fallow
Dhanbad	11.27	27.39	0.55	31.38	-	0.03	4.87	11.32	13.24
Jharia	2.11	6.45	-	75.46	0.27	0.23	2.26	9.04	4.16
Baliapur	26.86	16.08	3.13	20.79	1.20	0.57	6.46	12.58	12.33
Tundi	26.6	15.22	27.78	6.62	1.18	1.69	9.24	5.14	6.86
Nirsa	16.19	16.25	2.29	34.33	0.14	0.10	7.15	13.90	9.65
Baghmara	17.02	22.79	3.80	26.58	0.64	0.21	7.65	11.35	9.96
Topchanchi	25.85	11.58	13.55	11.62	-	0.21	6.70	19.35	11.19
Govindpur	31.44	12.86	3.27	13.09	0.75	1.54	6.10	19.52	11.43

Source : Calculated by author.

founding in hilly area in river beds of these Anchal. Jharia Anchal has 5.46% land under this category. Tundi is hilly where 9.6% area of total area covered as barren and uncultivated land. A compression of the barren land between 1980-81 and 2014-15, the proportion of barren land has decreased in Tundi, Baliapur and Baghmara Anchal due to better irrigation facilities provided to these Anchal during this period a considerable acreage of high barren land has been brought under cultivation.

X. LAND SOWN MORE THAN ONCE :

Dhanbad district has experienced a little change in the area sown more than once. Here cultivation depends upon monsoon only. So irrigation facilities are not available in after rainy season. Here and there tank and well is mainly source of irrigation so only 1032.01 acres in under land sown more than once. It is lying only .99 percentage of total area of the district.

REFERENCE :

1. Graham, E. H. : Natural Principals of Land Use, p. 97
2. Stamp, L.D. : The Land of Britain : It Use and Misuse, London
3. Stamp L.D. (1952) : Our Underdevelopment World, pp 158-59
4. Kohnke, L.D. (1967) : Soil Conservation, New York, pp. 152-54
5. Mandal, R.B. (1982) : Land Utilization Theory and Practise.
6. Ibid, pp. 156-57
7. Nath, V. (1953): Land Utilization in India, Vol.-1, p. 428
8. Ahmad, E. (1954) : Geographical Essays on India, Patna, pp. 15-16
9. SINHA, R.N.P. (1959) : 'Study of land use in a canal irrigated area of patna district' Unpublished Ph.D Thesis Patna University, pp. 177-79

4

Chapter A Study of the Sample Villages

In the previous chapter the selection of the sample villages have been done by the systematic purposive sampling method. The villages are widespread throughout the region. A detailed study of the land use of these sample villages is giving one by one in the following chapter.

BANGARO :

Bangaro is a hilly village of Tundi Anchal. It is 15 km west of Tundi and about 20 km north of Rajganj. It lies north of Tundi reserve forest. Bangaro lie on 23°58' N Latitude and 86°20' E Longitude. It is 20 km north of the national highway. The village is far away from the railway line. A small stream Gandua flows at five km west from the village. With the area of 542 acres the village has a population of 694 as per 2011 census.

Physiographically Bangaro is situated on higher land with a height of about 300 meters from the sea level. The general slope of the village is from south to north. The main physiographic landmark, River Gandua is in west. Tundi hill in the south and river Barakar in the east. The metalled road serves the village.

Between 1961 and 2011 the population of the village increased by 71.03 percentage. As regards literacy 59.11 percent of population is literate.

Out of the total population 78.5 percent is non-worker while all workers taken together accounts for only 21.55 percent. Among total population 3.9 percent are cultivator and 15.6 percent are agriculture labour this shows that the number of cultivator is less than that of agricultural labour.

Table - 4.1
POPULATION OF VILLAGE BANGARO

DESCRIPTION	POPULATION					VARIATION OF POPULATION			
	1961	1971	1981	1991	2011	1961-71	1971-81	1981-91	1961-2011
TOTAL POPULATION	201	252	317	397	694	20.91	20.50	20.15	71.03
LITERACY	16	26	75	112	413	-	-	-	-
PERCENTAGE OF LITERACY	7.86	10.31	23.65	28.11	59.51	-	-	-	-
DENSITY OF POPULATION (PER ACRES)	0.37	0.46	0.58	0.73	1.28	-	-	-	-
PER CAPITA LAND	2.70	2.15	1.71	1.36	0.78	-	-	-	-

Source: calculated by the author

LAND USE :

Out of the total area of 542 acres, 187 acres is culturable. This accounts for 34.55% of the total area. Thus 65.46% of total land is uncultivated out which the lion's share 190.20 acres goes to forest. Which is 35.09% of the total area. The forest is situated to southern portion, which is a part of Tundi reserve forest. Different types of trees found here. Sal is important among them. Forest area is comparatively higher land. Next comes waste land, which account for about 18.46 percent of the area it is situated in the middle portion of the village. It is relatively upland than net sown area. Here soil is less fertile and an irrigation facility is not available. Next comes not for cultivable lands, which cover 11.91% of the area that includes settlement, road, educational institutions and places of worship. The settlement is in linear pattern situated on both side of road. The details of the land use are shown in the Table no. 4.2

Table - 4.2
Land use Pattern of Bangaro (2015)

Sl. No.	Classification of land	Area in acres	Percentage of the total area of the land
1.	Forest	190.20	35.09
2.	Not available for cultivation	64.53	11.91
3.	Cultivation waste	100.00	18.46
4.	Fallow land	56.10	10.35
5.	Net area sown	131.10	24.19
	Total area of the village	**541.93**	**100.00**

Source: Calculated by the author

The area under cultivation is fairly low land where soil is most suitable. Step farming is practiced in this village as it is situated among a few hills. The facilities of irrigation are not adequate out of the total cultivated land about 10% is irrigated. Main source of irrigation is well and tank. The

Table - 4.3

Occupational Structure of the Working Force of Village Bangaro

Occupational Structure	Population					Percent of the total Population					Percent of total workers				
	1961	1971	1981	1991	2011	1961	1971	1981	1991	2011	1961	1971	1981	1991	2011
Cultivators	110	29	27	119	27	54.72	11.50	8.51	29.97	3.9	86.6	34.32	31.03	70.41	18.12
Agricultural labourers	14	50	53	50	108	6.96	19.84	16.71	12.60	15.6	11.02	59.52	60.91	70.41	18.12
Household industry	-	-	-	-	-	-	-	-	-	-	-	-	-	-	-
Marginal workers	-	-	-	-	0.10	-	-	-	-	1.4	-	-	-	-	6.71
Other workers	3	5	7	-	04	1.49	1.98	2.21	-	0.6	2.36	5.96	8.06	-	2.73
Total workers	127	84	87	169	149	-	-	-	-	1.4	100	100	100	100	100
Non workers	74	168	230	228	545	36.81	66.68	72.57	57.43	78.5					

Source: calculated by the author

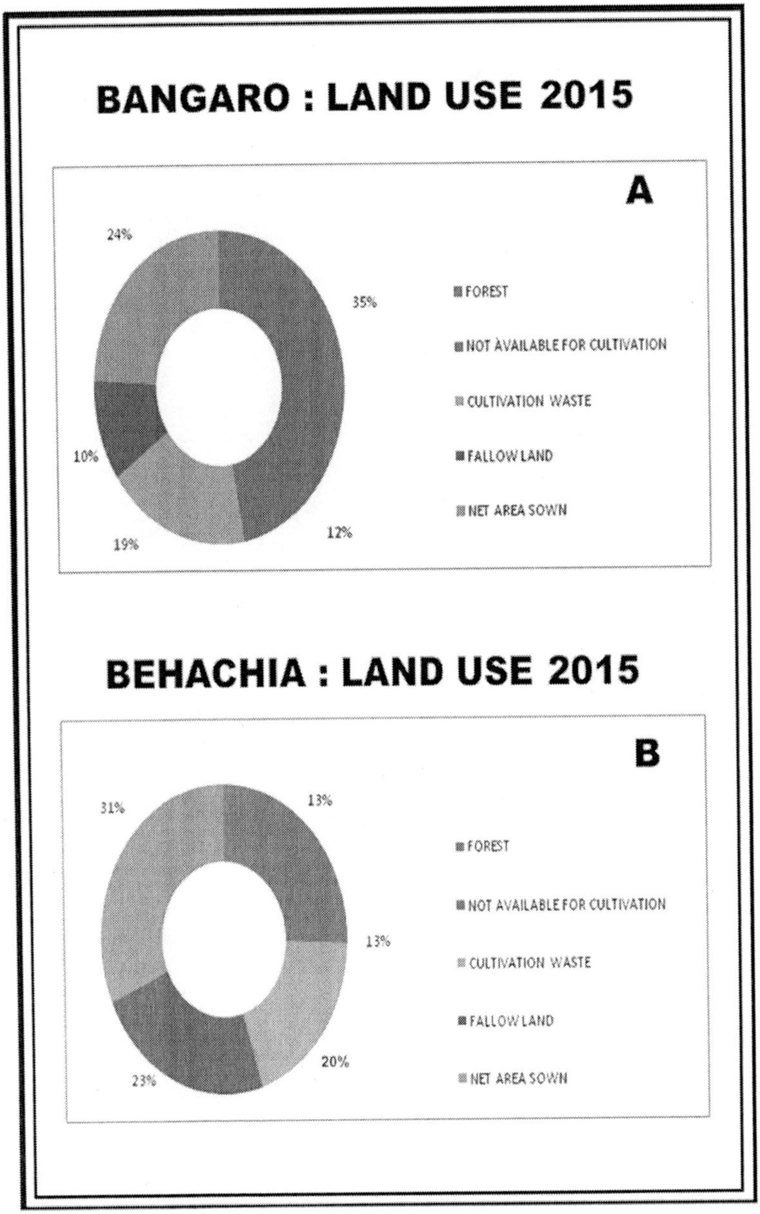

Fig. 4.1

Source : The Map Prepared by Author

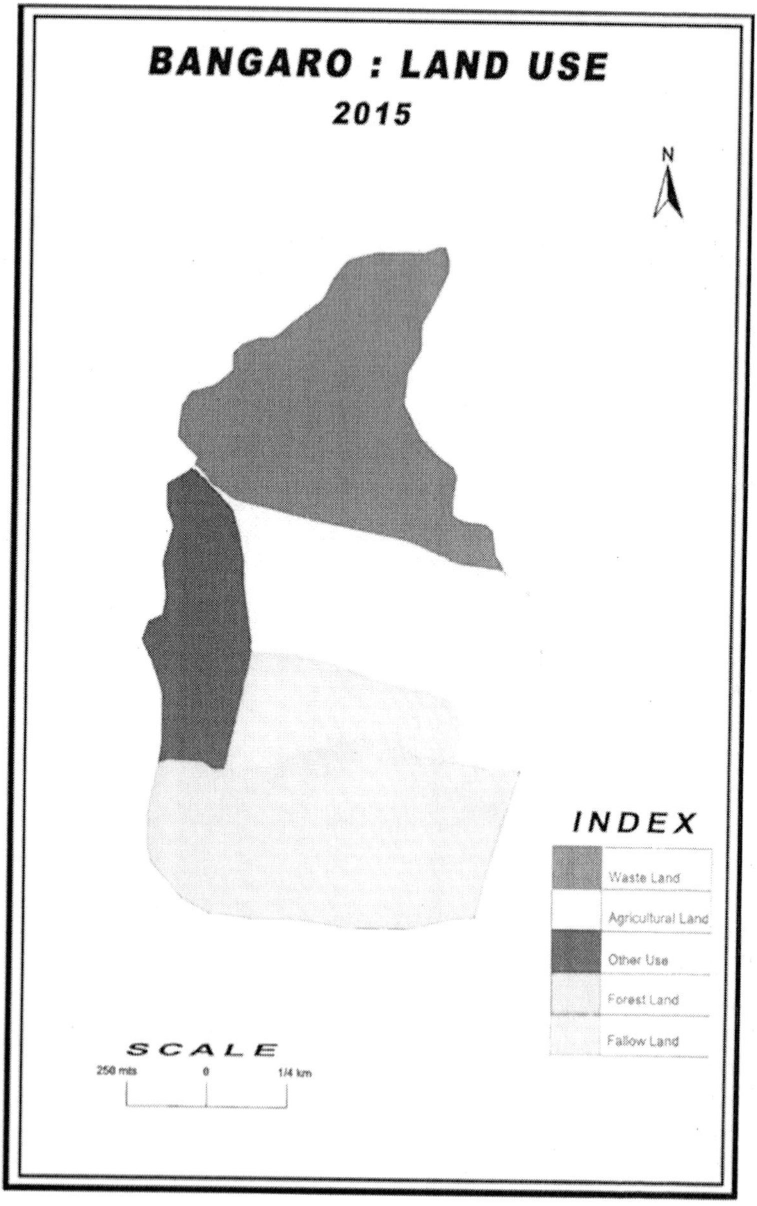

Fig. 4.2

Source : The Map Prepared by Author

cultivated land divided into two group (i) Fallow land (ii) the net area sown. About 56.10 acres area was under fallow land in 2015 where no types of crops were practiced. Due to lack of rainfall only 131.10 acres area is under net sown area, which shares 24.19% of the total area of village. About 90% of the total cultivated area is under single cropping while only 10% is under double cropping where irrigation facilities are available. Double cropping land is found in the northeast. Mainly two types of crops are grown here according to the season of harvest they are Aghani and Bhadai. Aghani is most important crop of the village covering about 80% to 85% of the total gross area sown of the village. Only two percent of the land is given to Rabi and about 5% to Bhadai, Aghani crop are paddy vegetables and sugarcane. Bhadai crops are Maize Kurthi, Marua and Urad. Rabi crops are wheat and potato.

BEHACHIA :

The village Behachia situated on 23057' north Latitude and 86020' E longitude. It falls in Govindpur Anchal and is located towards 20 km northwest of Govindpur town. It is bounded by Tundi reserve forest in the north; Tilaiya-Harladih road in the west, origin of river Katri is in the east of the village. This village has an area of 517 acres and population of 788 as per 2011 census. Thus the density of population of the village is 1.02 person per acre. Physiographically the village is situated on hilly portion of the district with the height of 280 meter from the sea level. The general slope of the village is from north west to south east. The higher land is situated to the north of the village from where the land slopes to the south east. Where lowest land is to be found. River Katri origins from Tundi reserve forest, which forms the eastern boundary of the village. Erosional fashion forms the waste land.

Out of the total population of 788, about 56 percent are literate. Non-workers constitute about 60 percent of the total population. Cultivator accounts for about 13 percent and agricultural labour 4% of the total population. Only about

Table - 4.4
Population of Village Behachia

DESCRIPTION	POPULATION					VARIATION OF POPULATION			
	1961	1971	1981	1991	2011	1961-71	1971-81	1981-91	1961-2011
TOTAL POPULATION	339	418	475	529	788	18.89	12.0	10.20	56.97
LITERACY	38	50	111	157	440	-	-	-	-
PERCENTAGE OF LITERACY	11.20	11.96	23.36	29.67	55.83	-	-	-	-
DENSITY OF POPULATION (PER ACRES)	0.65	0.80	0.91	1.02	1.52	-	-	-	-
PER CAPITA LAND	1.52	1.23	1.08	0.93	0.65	-	-	-	-

Source: Calculated by the author

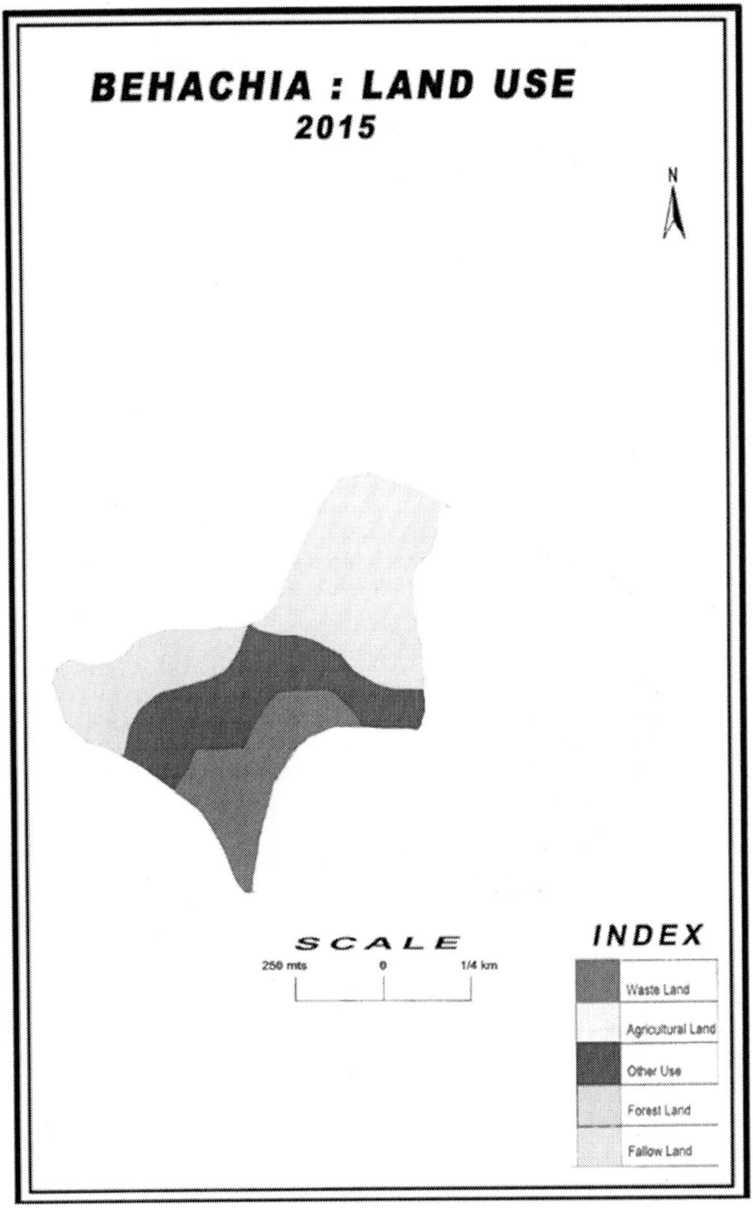

Fig. 4.3

Source : The Map Prepared by Author

7% of the population is other worker.

The population of the village has increased by 57 percent during 1961-2011. It is important to note that the variation in population has not been uniform during these three decades. It has experienced the highest growth of population during 1961-71 when the population has increased by about 19%. The highest rate of growth during 1961-71 is attributes to the low rate of mortality due to Medicare. After that the growth rate has slowed down due to a check on the rate of fertility due to family planning measures. (Table no. 4.5)

Table - 4.5
Land use pattern of Behachia

Sl. No.	Classification of land	Area in acres	Percentage of the total area of the land
1.	Forest	50.32	9.74
2.	Not available for cultivation	50.74	9.80
3.	Cultivation waste	78.15	15.12
4.	Fallow land	92.33	17.86
5.	Net area sown	245.40	47.48
	Total area of the village	516.94	100.00

Source: Calculated by the author

LAND USE :

The village Behachia occupies 517 acres of land out of which 338 acres are devoted to cultivation. From agriculture point of view the village is important as more than 65% of its land use cultivated. Only about 35% left for non-agricultural use. In which about 10% is for forest and same percent is under non available for cultivation among them settlement, water bodies, road, orchard and educational centres covers about 15% land is Culturable waste. Culturable waste land is upland and now it is not suitable for cultivation. Of the total cultivated area 96% is unirrigated and only 4% is irrigated. As in other villages three types of cropping is practiced here.

Table - 4.6

Occupational Structure of the Working Force of Village Behachia

Occupational Structure	Population					Percent of the total Population					Percent of total workers				
	1961	1971	1981	1991	2011	1961	1971	1981	1991	2011	1961	1971	1981	1991	2011
Cultivators	178	116	148	194	101	92.50	27.75	31.16	36.69	12.81	83.96	76.82	68.81	82.91	31.86
Agricultural labourers	15	20	28	22	16	4.42	4.78	5.89	4.16	2.0	7.08	13.24	12.96	9.40	5.7
Household industry	19	14	24	10	41	5.52	3.45	5.05	1.89	5.20	8.96	9.27	11.11	4.27	12.93
Marginal workers	-	-	27	-	53	-	-	5.68	-	6.72	-	-	-	-	16.7
Other workers	-	1	16	8	106	-	0.23	3.37	1.51	13.45	-	0.66	7.41	3.42	33.43
Total workers	212	151	216	234	317	-	-	-	-	-	100	100	100	100	
Non workers	127	267	232	295	471	37.46	30.38	48.84	55.76	59.77					

Source: calculated by the author

Percentage of area sown more than once is about 5%. The more fertile land of the village situated near by the river is given to double cropping. Where wheat is cultivated in Rabi season. Single cropping practiced in most of sown area. Among Aghani crop paddy is cultivated almost Maize and Pulses are the important Bhadai crops. Table no. 6.5 shows the detailed land use of the village. About 35% of land use accounts for non-agricultural use. Among non-agricultural use about 10% is covered by forest. Not available for cultivation land occupied about 15% in which, settlement, religious places, educational institutions and road is situated.

TETULIA :

Tetulia is located in Nirsa Anchal. It is situated on 23°48' N latitude and 86°37' E longitude. It is only 5 miles away from Nirsa town, the Anchal headquarter. The village is situated 20 km Northeast of Dhanbad and 12 km east of Govindpur. It is situated on GT road (NH 2). The village is served by river Khudia that is a tributary of river Damodar.

Tetulia is located on a relatively intermediate land. Where the land slopes toward south west. The village is situated at the height of 260 meter from the sea level. The relatively upper parts of the village lies to north and North east. River Khudia enters into the village from North West side and runs parallel to its western boundary. In rainy season the river has much water.

Tetulia is spread over an area of 485 acres the total population of the village as per 2011 census is 1666. The density of population is 3.07 person per acre. The percentage of literate is about 60% of the total population (table no. 4.7)

Table - 4.7
Population of Village Tetulia

DESCRIPTION	POPULATION					VARIATION OF POPULATION			
	1961	1971	1981	1991	2011	1961-71	1971-81	1981-91	1961-2011
TOTAL POPULATION	363	405	767	782	1666	10.37	47.19	48.45	78.21
LITERACY	42	112	211	369	989	-	-	-	-
PERCENTAGE OF LITERACY	11.57	27.65	27.50	247.18	59.36	-	-	-	-
DENSITYOF POPULATION (PER ACRES)	0.74	0.83	1.58	1.61	3.07	-	-	-	-
PER CAPITA LAND	1.33	1.99	0.63	0.62	0.32	-	-	-	-

Source: Calculated by the author

Table - 4.8
Occupational Structure of the Working Force of Tetulia

Occupational Structure	Population					Percent of the total Population					Percent of total workers				
	1961	1971	1981	1991	2011	1961	1971	1981	1991	2011	1961	1971	1981	1991	2011
Cultivators	135	33	55	90	83	37.19	81.15	7.17	11.50	4.98	74.59	29.46	17.26	38.62	12.27
Agricultural labourers	1	37	27	11	180	0.27	9.13	3.52	1.41	10.80	0.55	33.03	8.46	4.72	26.62
Household industry	6	18	20	25	5	1.65	4.44	2.61	3.20	0.3	3.31	16.07	6.26	10.73	0.73
Marginal workers	-	-	-	-	274	-	-	-	-	16.44	-	-	-	-	40.53
Other workers	39	24	217	107	134	10.74	5.92	28.29	3.68	8.0	21.55	21.44	68.02	45.93	16.82
Total workers	181	112	319	233	676	?	?	?	?	40.58	100	100	100	100	100
Non workers	182	293	448	550	990	50.13	72.35	58.41	70.33	59.42					

Source: Calculated by the author

As regards occupational structure of population 60% of the of the people are non-worker. Other includes cultivators, agricultural labour, other worker and marginal worker. Which constitutes 5%, 11%, 8% and 16% respectively. (Table no. 4.8)

LAND USE :

Tetulia is an agricultural village but the facilities of agriculture are inadequate. As such the only source of irrigation is river which dries up in summer. Due to this the about 10% of the total cultivated area is irrigated and 90% is left on the mercy of monsoon. About 52% of the total land is cultivated but of the gross cultivated area 44% is under fallow land. Out of 485 acres only 175 acres is covered by net sown area. Which is 36% of the total area of the village. The pattern of cropping in the village is similar to that of other village.

Double cropping is practiced around the settlement and also toward the southern part of the village. About 7% of the gross cultivated area is given to double cropping. Single cropping is practiced in rainy season. Total fallow land is comparatively upland where lack of irrigation facility is main cause of fallow.

Aghani crops occupied about 85% of the gross area sown, Bhadai crop share about 9% and 2% share to the Rabi crop. Paddy is main Aghani crop. Maize and Vegetable are Bhadai crop. Wheat and potato are Rabi crop.

About 48% of the total area of Tetulia in uncultivated, out of which more than 27% is not available for cultivation. Road, settlement, riverbed, lie in the group. About 20% land available for cultivable wasteland. This type of land is mainly situated in the northern portion of the village and is respectively upland. Where soil is less fertile. Waste land also found near riverside. The land use of Tetulia is shown in the Figure no. 4.4 (A)

Table - 4.9
Land use Pattern of village Tetulia

Sl. No.	Classification of land	Area in acres	Percentage of the total area of the land
1.	Forest	190.20	35.09
2.	Not available for cultivation	64.53	11.91
3.	Cultivation waste	100.00	18.46
4.	Fallow land	56.10	10.35
5.	Net area sown	131.10	24.19
	Total area of the village	541.93	100.00

Source: Calculated by the author

PAHARPUR :

Paharpur is the representatives village of Baliapur Anchal. It is situated by the side of Baliapur Govindpur road and lie on 23°43' north latitude and 86°32' E longitude. It is 5 km North West of Baliapur town, the Anchal headquarter. It is half a kilometre north of Jharia-Baliapur road which is metalled and accessible in all seasons. Pradhan Khanta is the nearest railway station, which is about 7 km north of the village.

The village is situated on a comparatively lowland with a height of 180 meter above the sea level. The general slop is from south to north and the highest land lie towards the south.

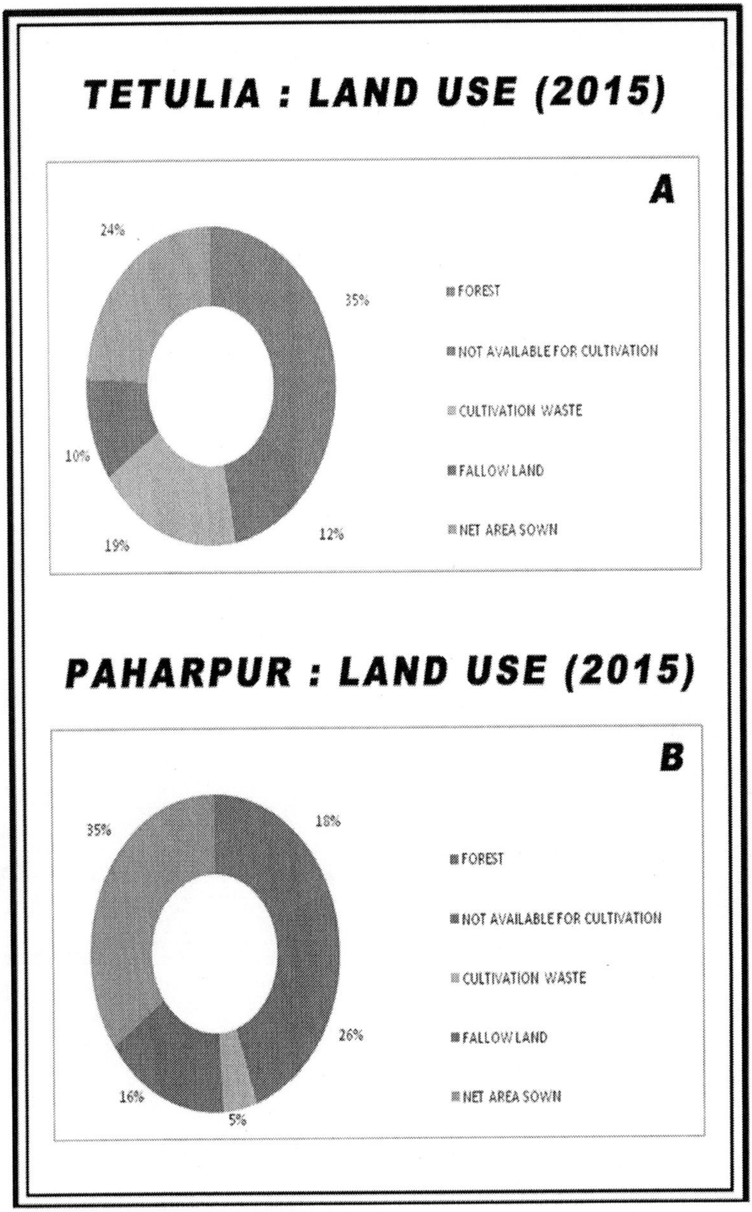

Fig. 4.4

Source : The Map Prepared by Author

Table - 4.10
Population of village Paharpur

DESCRIPTION	POPULATION					VARIATION OF POPULATION			
	1961	1971	1981	1991	2011	1961-71	1971-81	1981-91	1961-2011
TOTAL POPULATION	749	863	1164	1488	1819	13.20	25.85	21.77	58.82
LITERACY	77	162	280	468	1165	-	-	-	-
PERCENTAGE OF LITERACY	10.28	18.79	24.09	33.45	64.04	-	-	-	-
DENSITYOF POPULATION (PER ACRES)	1.41	1.62	2.19	2.80	3.43	-	-	-	-
PER CAPITA LAND	0.70	0.61	0.45	0.35	0.29	-	-	-	-

Source: Calculated by the author

It is drained by small stream which forms the north and north eastern border. The stream also dries up after the rainy season. Paharpur is spread over an area of 530 acres and a population 1819 as per 2011 census.

Thus, the density of population is 3.43 person per acres (tables no. 4.10). About 64% of its population is literate. About 32.61% of the population are workers. Among the working force 4.7% are cultivators, 7% other workers and about 4% are agricultural laboures. Other worker including trades, professionals and service holders. As regards growth of population the village has experience at the rate of 21% during decades 1981-91. During 1961-2011 the village has experience a growth in its population 58.82%.

LAND USE :

Out of the total area of the village about 270 acres is cultivated which works out to be 51%. Out of cultivated land 10% is irrigated and the rest is dependent on rain water, of the total cultivated area. 20% is given to double cropping which is practiced on Bari land near the river side. It lies north and eastern side of the village. Rest land is devoted to single cropping.

As in other village Bhadai and aghani are the most important harvests which is done in about 80% of the gross cultivated land. Paddy and maize are main crops. Wheat and potato are the important rabi crops. Vegetable are sown in Bhadai, aghani and garma season.

About 49% of the land of Paharpur is uncultivated in which 18.39% covered by forest. The forest has a few no. of trees. Now it is hip of stone only where number of stone crusher is situated. Among non-agriculture use of land settlement covers by far the largest area. It is spread over 26% of the land of the land of the village. It occupied the higher land at the center. Culturable wasteland covers 4.41% of the total land. A vast tract of land lies in the bed of the river.

Table - 4.11
Occupational Structure of the Working Force of Village Paharpur

Occupational Structure	Population					Percent of the total Population					Percent of total workers				
	1961	1971	1981	1991	2011	1961	1971	1981	1991	2011	1961	1971	1981	1991	2011
Cultivators	321	151	88	93	67	42.80	20.97	7.56	6.25	3.68	81.06	70.89	26.91	27.51	11.29
Agricultural labourers	1	4	48	54	108	0.13	0.46	4.13	3.63	5.93	0.25	1.88	14.68	15.89	18.21
Household industry	-	-	-	-	01	-	-	-	-	0.05	-	-	-	-	0.19
Marginal workers	-	-	-	-	288	-	-	-	-	15.83	-	-	-	-	48.56
Other workers	74	58	191	191	129	9.88	6.72	16.40	8.80	7.0	18.67	21.23	58.40	56.51	21.75
Total workers	396	213	327	388	593	-	-	-	-	32.61	100	100	100	100	100
Non workers	353	650	767	1150	1226	47.12	75.31	65.89	77.28	67.39					

Source: Calculated by the author

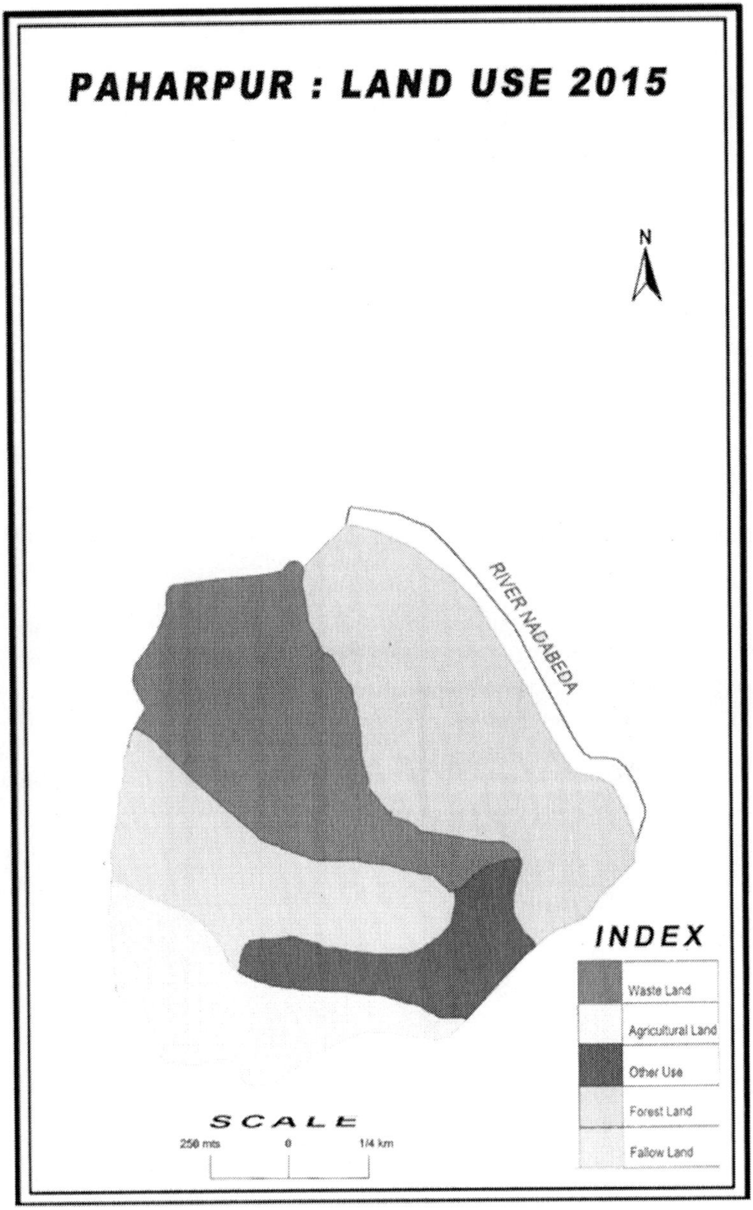

Fig. 4.5

Source : The Map Prepared by Author

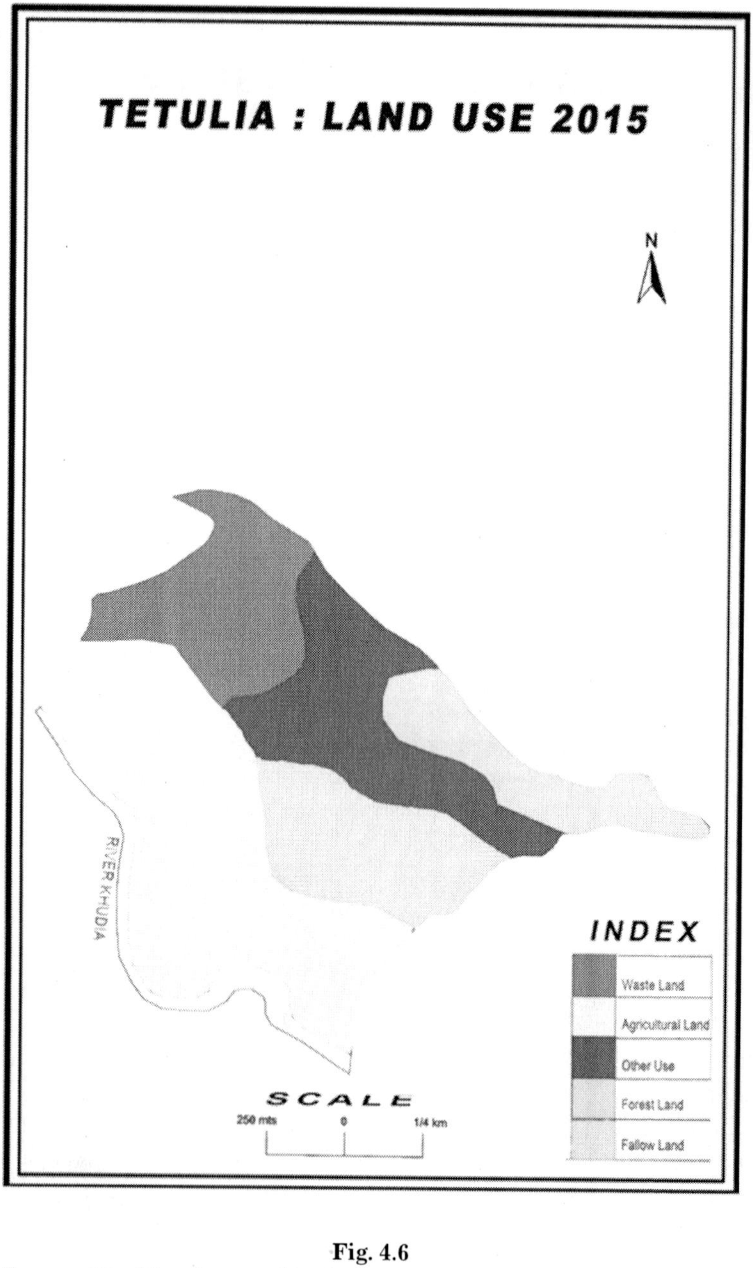

Fig. 4.6

Source : The Map Prepared by Author

Table - 4.12
Land use Pattern of village Paharpur

Sl. No.	Classification of land	Area in acres	Percentage of the total area of the land
1.	Forest	97.44	18.39
2.	Not available for cultivation	139.34	26.30
3.	Cultivation waste	23.39	4.41
4.	Fallow land	85.10	16.06
5.	Net area sown	184.50	34.84
	Total area of the village	529.77	100.00

Source: Calculated by the author

SALDAHA :

Lying on 23°51' N latitude and 86°20' E longitude, Saldaha is situated on the northern portion of Baghmara Anchal. It is hardly 2 km. south east of Rajganj. It is connected with katras by metalled road. Through Rajganj it is well connected with Dhanbad town. The nearest railway station Tetulmari is about 5 km away from the village.

Saldaha is located on intermediate land of Dhanbad district. Its general height above sea level is 220 meter. The general slope of the village is from north to south, although the comparatively higher land is situated at the center part of the village. A small nala forms the eastern and southern boundary of the village, but this nala is almost dead and is filled with water only in the rainy season. Good quality of soil found in the law land area while Balsundary soil is also found near the nala.

Saldaha is a small village spread over an area of 96 acres with population of 334 person according to 2011 census. The density of population is 1.4 person per acres. About 64% of

Table - 4.13
POPULATION OF SALDAHA

DESCRIPTION	POPULATION					VARIATION OF POPULATION			
	1961	1971	1981	1991	2011	1961-71	1971-81	1981-91	1961-2011
TOTAL POPULATION	103	110	134	238	334	6.36	17.9	43.69	69.16
LITERACY	10	14	26	54	210	-	-	-	-
PERCENTAGE OF LITERACY	9.70	10.72	19.40	22.68	63.87	-	-	-	-
DENSITYOF POPULATION (PER ACRES)	0.43	0.46	0.56	1.00	1.40	-	-	-	-
PER CAPITA LAND	1.52	1.23	1.08	0.93	0.70	-	-	-	-

Source: Calculated by the author

Table - 4.14
Occupational Structure of the Working Force of Village Saldaha

Occupational Structure	Population					Percent of the total Population					Percent of total workers				
	1961	1971	1981	1991	2011	1961	1971	1981	1991	2011	1961	1971	1981	1991	2011
Cultivators	51	33	27	17	15	49.51	30.00	16.42	7.14	4.49	91.18	91.66	56.41	20.99	11.11
Agricultural labourers	-	1	1	8	50	-	0.91	0.29	3.36	14.97	-	2.74	2.57	9.88	37.03
Household industry	-	-	-	-	-	-	-	-	-	-	-	-	-	-	-
Marginal workers	-	-	-	10	19	-	-	-	4.20	5.68	-	-	-	12.34	14.09
Other workers	5	2	16	46	51	4.15	1.82	4.79	19.31	15.26	8.92	5.56	41.02	56.79	37.77
Total workers	56	36	39	81	135	-	-	-	-	40.42	100	100	100	100	100
Non workers	47	74	95	157	199	45.63	67.27	28.44	65.96	59.58					

Source: Calculated by the author

the population is literate. As regards decennial rate of growth of population it has rapidly increased during 1981-91 as compared to 1971-81. During 1981-91 the population has grown at the rate 43.69%. While during 1961-71, it has only 6.36%. The high rate of growth during recent years may be attributed to the developed Medicare and other living condition. During this period 3 hard coke plant was established which attracted a number of worker with their families to settle here. Coming to occupational structure of population Saldaha has a high percentage 40% of worker to the total population only 60% of the people are non worker. Cultivator account for 4% of the total population while other workers share 15%. This shows the impact of industrialization. About 15% of the people is agricultural labourer.

LAND USE :

As a typical agricultural village 84% of the land of Saldaha is cultivated while only 16% is uncultivated. Of 237 acres of land 200 acres is cultivated and only 37 acres is uncultivated. But of the cultivated land only 3% is irrigated. This is possible only due to availability of water by small nala and two tanks. Lack of irrigation facilities reflected in the cropping pattern. Almost area is under single cropping. Double cropping area is situated to the southern part in a long belt along the river and also around the settlement at the center.

Aghani is the most important harvest of the village and covers about 70% of the gross cultivated area. Paddy and potato are main crops. Wheat, Arhar and Sugarcane are under Rabi Crop. Maize is only Bhadai crop grown near settlement.

About 16% of the total area of the village is uncultivated. Among non-agricultural use of land settlement road and factory covers 9.47% of the land of the village. It occupied the higher land at the center. About 6% land covered by Culturable waste land among them mostly part is up land which is erosioned by rain water.

SALDAHA : LAND USE (2015)

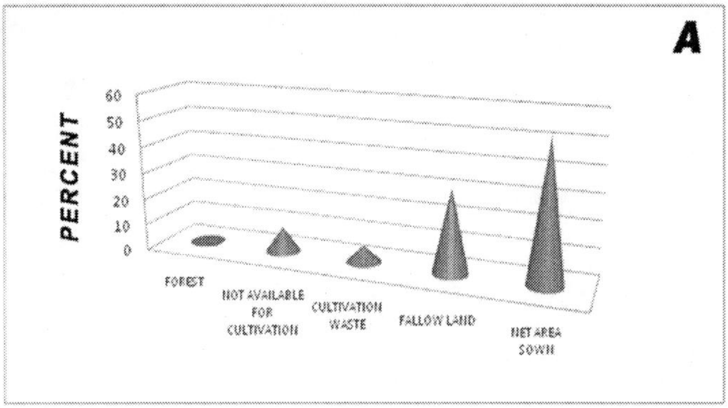

KHARNI : LAND USE (2015)

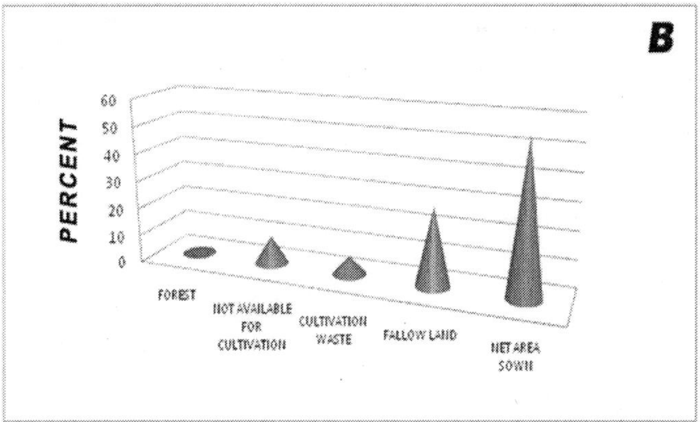

Fig. 4.7

Source : The Map Prepared by Author

Fig. 4.8

Source : The Map Prepared by Author

Table - 4.15
LAND USE PATTERN OF VILLAGE SALDAHA

Sl. No.	Classification of land	Area in acres	Percentage of the total area of the land
1.	Forest	-	-
2.	Not available for cultivation	22.46	9.47
3.	Cultivation waste	14.99	6.32
4.	Fallow land	75.09	31.66
5.	Net area sown	124.55	52.55
	Total area of the village	237.07	100.00

Source: Calculated by the author

KHARNI :

Kharni is situated in Topchanchi Anchal. It lies on 24°51' N latitude and 86°17' E longitude. It is an interior village away from the metalled road and railway line it is about 3 km west of Rajganj-katras road. The village is about 20 km south east of Topchanchi, the Anchal head quarter.

Kharni is a relatively intermediates land with a height of about 220 meter above the sea level. The village slops to the south which is evident from the flow of the water.

With an area of 457 acres and population of 951, the density of population of the village is low 2.08 person per acres. About 49% of the population are literate. (Table No. 4.16) the working force constitute about 40% of the population. Out of which 2.62% are cultivators, 8.20% are agricultural labourer. This speaks of the semi agricultural economy of the village. About 19% of the people are engaged in other work. Which include trade and commerce, government service and other profession. The percentage of non worker is only about 60%. As regard growth of population

Table - 4.16
Population of village Kharni

DESCRIPTION	POPULATION					VARIATION OF POPULATION			
	1961	1971	1981	1991	2011	1961-71	1971-81	1981-91	1961-2011
TOTAL POPULATION	406	421	466	641	951	3.56	9.65	27.30	57.30
LITERACY	30	51	109	182	465	-	-	-	-
PERCENTAGE OF LITERACY	7.38	12.11	23.39	28.39	48.89	-	-	-	-
DENSITYOF POPULATION (PER ACRES)	0.88	0.92	1.01	1.40	2.08	-	-	-	-
PER CAPITA LAND	1.12	1.08	0.98	0.71	0.48	-	-	-	-

Source: Calculated by the author

the village has experienced a growth of about 57% is 50 year between 1961 and 2011. The decennial rate of growth has not uniform during 1981-91 the rate of growth was 27.30%. While during 1961-71 and 1971-81 was 3.56% and 9.65% respectively. (Table no. 4.16) the highest rates of growth during 1981-91 attributed to low rate of mortality, due to Medicare and don't check by family planning.

LAND USE :

As in other agricultural villages about 80% of land is cultivated out of which 15% is irrigated. Main source irrigation is well and tank big tank are located in the southern portion. 6 wells are situated on the middle area of the village it is built by govt. irrigation purpose. In spite of it a large tract of agricultural land is unirrigated. The main feature of the agricultural of the village is the single cropping only. About 20% of land is under double cropping. It is situated near by settlement and southern part of the village. Where tanks are located. As regards harvest of the village Aghani and Bhadai dominate the seasonal harvest. Covering 80% of the gross cultivated land. Aghani crops include paddy, potato, sugarcane and vegetable, Maize kurthi etc. are the Bhadai crops cover 65% of the land. Under Rabi harvest wheat and vegetable are the main crops. Recently cropping vegetable, become popular. Cultivator used to send own product to nearer urban center and earn more money.

Out of 457 acres land only 78 acres is given to non agricultural use. This is 17% of the total area. Of the uncultivated land 6.84% is under settlement, educational institution and worship place. The settlement is clustered at one place towards the mid western part of the village. It covers the highest land of the village. It is surrounded by orchards and sacred place. About 6.84% of land of total land is Culturable waste land.

Table - 4.17

Occupational Structure of the Working Force of Kharni

Occupational Structure	Population					Percent of the total Population					Percent of total workers				
	1961	1971	1981	1991	2011	1961	1971	1981	1991	2011	1961	1971	1981	1991	2011
Cultivators	171	75	44	44	25	42.11	17.81	9.44	6.86	2.62	74.34	55.35	33.85	15.22	6.63
Agricultural labourers	2	51	31	20	78	0.49	12.11	6.65	18.72	8.20	0.86	36.95	23.85	41.52	20.68
Household industry	1	-	-	-	-	0.25	-	-	-	-	0.43	-	-	-	-
Marginal workers	-	-	-	-	87	-	-	-	-	9.14	-	-	-	-	23.07
Other workers	56	12	56	125	189	13.798	2.85	12.01	19.50	19.87	24.37	8.69	42.40	43.26	50.12
Total workers	230	138	130	289	377	-	-	-	-	39.65	100	100	100	100	
Non workers	176	283	331	352	574	43.34	67.22	71.03	54.91	60.35					

Source: Calculated by the author

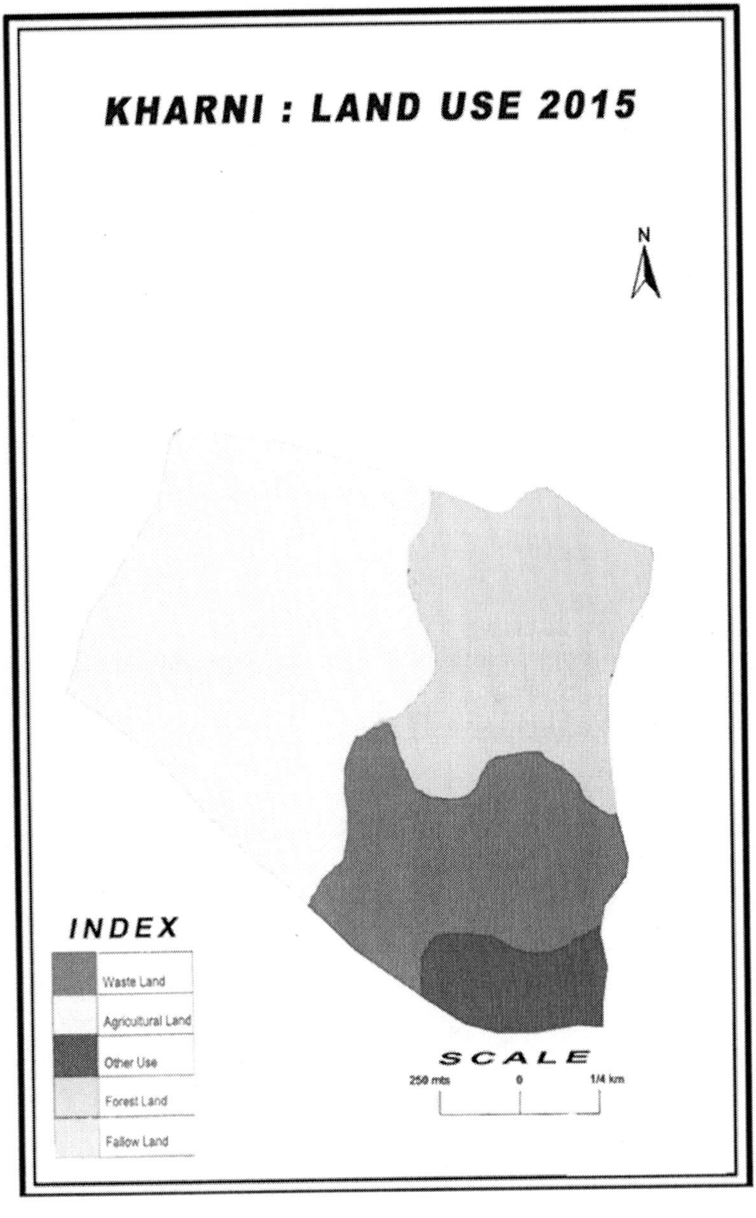

Fig. 4.9

Source : The Map Prepared by Author

Table - 4.18
LAND USE PATTERN OF VILLAGE KHARNI

Sl. No.	Classification of land	Area in acres	Percentage of the total area of the land
1.	Forest	-	-
2.	Not available for cultivation	46.89	10.26
3.	Cultivation waste	31.20	6.84
4.	Fallow land	129.03	28.23
5.	Net area sown	249.86	54.67
	Total area of the village	456.98	100.00

Source: Calculated by the author

SIALGUDARI :

Sialgudari is situated just by the side of putki town. It is actually the suburban part of putki. It lies on 23°46' N latitude and 86°21' E longitude. It is just west of putki town and situated along Dhanbad Bokaro road line. It is 8 km. West of Dhanbad town. The village has an ideal location in respect of transport and communication as it served by important road line.

The village is situated at a lower height compared to those discussed earlier. Its average height is 200 meter above the sea level, which is shown in the nearby contour line. This village also slopes from south west to north east.

Sialgudari is spread over an area of 253 acres where 5450 person live according to 2011 census. Thus the density of population is 21.45 person per acres. A glance at the census figure of 1961, 1971, 1981, 1991 and 2011 show that the population has increased per decade. The population has grown by 64% during 1961-2011 and by 91.24% during 1971-81, 29.51% during 1981-91. Thus the overall increase during

Table - 4.19

POPULATION OF VILLAGE SIALGUDARI

DESCRIPTION	POPULATION					VARIATION OF POPULATION			
	1961	1971	1981	1991	2011	1961-71	1971-81	1981-91	1961-2011
TOTAL POPULATION	477	1327	2591	3676	5450	64.0	48.86	29.51	91.24
LITERACY	165	406	810	1499	3740	-	-	-	-
PERCENTAGE OF LITERACY	34.59	30.64	31.26	40.77	68.62	-	-	-	-
DENSITYOF POPULATION (PER ACRES)	1.88	5.22	10.22	14.50	21.45	-	-	-	-
PER CAPITA LAND	0.53	0.19	0.09	0.06	0.04	-	-	-	-

Source: Calculated by the author

Table - 4.20
Occupational Structure of the Working Force of Sialgudari

Occupational Structure	Population					Percent of the total Population					Percent of total workers				
	1961	1971	1981	1991	2011	1961	1971	1981	1991	2011	1961	1971	1981	1991	2011
Cultivators	39	71	22	28	9	8.18	5.36	0.84	0.76	0.16	21.90	12.11	3.00	2.89	0.59
Agricultural labourers	13	72	5	6	60	2.73	5.43	0.19	0.16	1.10	7.30	12.94	0.68	0.61	3.99
Household industry	5	20	32	45	113	1.05	1.51	1.24	1.23	2.07	2.81	3.59	4.36	4.64	7.52
Marginal workers	-	-	-	25	110	-	-	-	0.68	20.36	-	-	-	2.58	73.95
Other workers	121	393	674	866	209	25.37	29.67	26.02	23.55	30.83	67.98	70.68	91.95	89.28	13.95
Total workers	178	556	733	970	1501	-	-	-	-	27.65	100	100	100	100	100
Non workers	299	769	1838	2633	3949	62.67	58.03	70.94	71.67	72.45					

Source: Calculated by the author

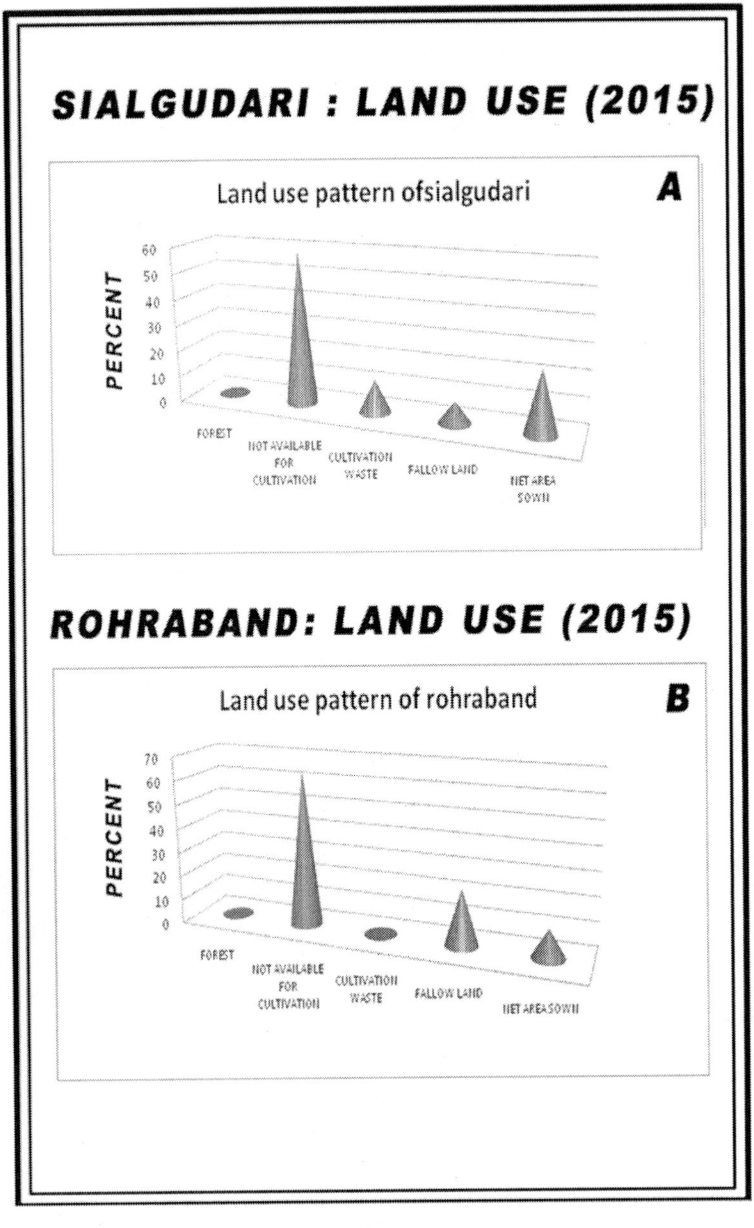

Fig. 4.10

Source : The Map Prepared by Author

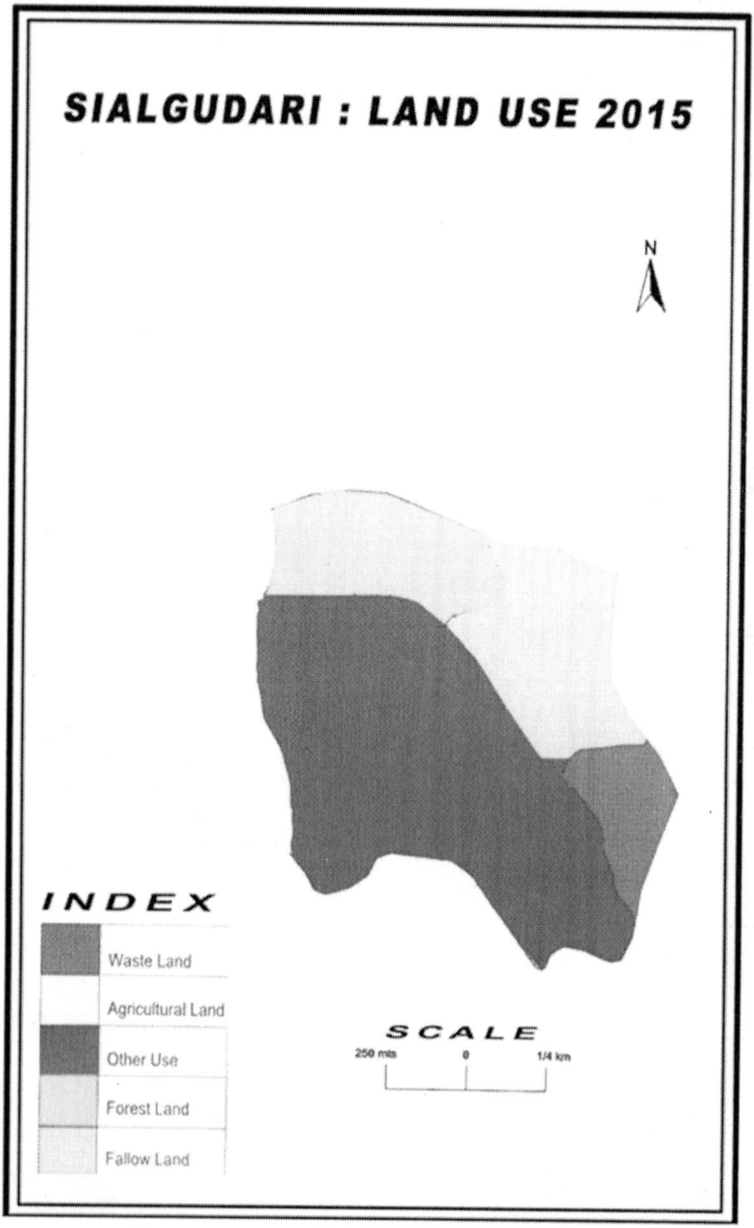

Fig. 4.11

Source : The Map Prepared by Author

1961-91 at the rate of 87.02%. (Table 4.19). Rapid increasement of population due to urbanization of the locality. Mostly people come here to do work in coal field mining. Munidh project one of the largest project of India is nearby.

LAND USE :

The general land use of the village shows that about 73% of its area is of the uncultivated land. Major share goes to non available for cultivation. Which cover settlement institution workshop, water bodies road etc. It is 60% of the total area of the village. (Table no.4.19). The settlement occupies the highest land to the north west of the village and both side road. Thus the settlement is of clustered type. The only water body of the village is a tank. Which is situated at the center of the village. Out of 253 acres about 32 acres occupy by Culturable waste land. it covers orchards also which is along road line.

About 50% of gross cultivation land is now fallow land. Net sown area occupied only 50 acres in which no irrigation facilities are available. Due to mining area water table is very low. So people get drinking water also by supply. Net sown area is a single cropping area where paddy is crop, as other villages. Vegetable are grown in Bhadai season. A little acre of land practiced in Bhadai harvest by maize.

Table - 4.21
Land use pattern of village Sialgudari

Sl. No.	Classification of land	Area in acres	Percentage of the total area of the land
1.	Forest	-	-
2.	Not available for cultivation	150.06	59.22
3.	Cultivation waste	32.24	12.74
4.	Fallow land	21.02	8.29

5.	Net area sown	50.06	19.75
	Total area of the village	253.38	100.00

Source: Calculated by the author

ROHRABAND :

Rohraband is situated on the southern border of Dhanbad district. It falls in Jharia Anchal it is the largest among all of the sample village studied. It occupies an area of 1153 acres and has a population of 8248 as per 2011 census. The village lie in urban area so it has post office, dispensary, and high school and telegraph office. It lies on 23°40'N latitude and 86°29'E longitude. It is situated on Jharia-Sindri road. Jharia-Sindri road is a state highway.

Rohraband is about 20 km southeast of Jharia and 10 km west of Sindri. Rohraband is located on coal mining belt. Its average height from sea level is 180 meter. The general slope of village is from North to South. But the land variations are also found here. The highest land village is to the north of state highway. Where settled area is located. The village touches by river Damodar also.

Over 1153 acres of land Rohraband 8248 person live. Thus the density of population is 32.60 person per acre. 89% of the population of the village is literate (Table no 4.22).

As regards occupational structure of population the village is different from other. Firstly the number of agricultural labour, cultivators and household industry worker are a few, about only 0.54 percent of the working force. Secondary the population of other worker is very high. About 29 percent of total population are engaged in other works, such as transport and communication, Govt. Services and other profession. It is due to the fact that village is a part of urban area. Rohraband has experienced a high rate of growth population during 1961-2011.when population has risen to 69% in 2011 over the figure of 1961. But the rate of growth has been uneven. The rate has been higher during 1971

Table - 4.22
Population of village Rohraband

DESCRIPTION	POPULATION					VARIATION OF POPULATION			
	1961	1971	1981	1991	2011	1961-71	1971-81	1981-91	1961-2011
TOTAL POPULATION	-	2756	4185	5548	8284	-	34.15	26.56	69.01
LITERACY	-	920	1955	3931	7380	-	-	-	-
PERCENTAGE OF LITERACY	-	33.39	46.71	70.85	89.47	-	-	-	-
DENSITYOF POPULATION (PER ACRES)	-	6.07	9.22	12.22	32.60	-	-	-	-
PER CAPITA LAND	-	0.16	0.10	0.8	0.13	-	-	-	-

Source: Calculated by the author

Table - 4.23
Occupational Structure of the Working Force of Rohraband

Occupational Structure	Population					Percent of the total Population					Percent of total workers				
	1961	1971	1981	1991	2011	1961	1971	1981	1991	2011	1961	1971	1981	1991	2011
Cultivators	-	-	40	30	22	-	-	0.95	0.54	0.13	-	-	3.08	1.78	0.90
Agricultural labourers	-	24	8	5	5	-	0.87	0.19	0.09	0.06	-	2.36	0.67	0.29	0.20
Household industry	-	5	9	12	48	-	0.18	0.21	0.22	0.58	-	0.49	0.69	0.71	1.94
Marginal workers	-	-	24	30	110	-	-	0.51	0.54	1.33	-	-	1.85	1.78	4.45
Other workers	-	987	1241	1610	2285	-	35.81	29.65	29.02	27.70	-	97.1	95.61	95.44	92.51
Total workers	-	1016	1298	1687	2470	-	-	-	-	29.94	100	100	100	100	100
Non workers	-	170	2863	3861	5778	-	6.17	68.41	69.60	70.06					

Source: Calculated by the author

Fig. 4.12

Source : The Map Prepared by Author

(34.15%) than that during 1981-91 (24.5%) (Table no. 4.22).

LAND USE :

About 90% of the area of the village is uncultivated and only 10% is cultivated. A few acres of area is irrigated. Most of the cultivated land faces vagaries of monsoon. Only single cropping is done. Paddy is main crop. Of the non-agricultural use of land open caste mining covers lion's share settlement also cover big portion and not available for cultivation.

Non cultivable land shows an increase due to the increasing demand of land for settlement especially urban settlement road, industrial establishment and mining.

Table - 4.24
Land use pattern of village Rohraband

Sl. No.	Classification of land	Area in acres	Percentage of the total area of the land
1.	Forest	-	-
2.	Not available for cultivation	749.52	65.01
3.	Cultivation waste	-	-
4.	Fallow land	269.05	23.34
5.	Net area sown	134.32	11.65
	Total area of the village	1152.91	100.00

Source: Calculated by the author

Table - 4.25
LAND USE AND WORKING FORCE OF THE STUDY AREA

Name of the Village	Area in acres	Population 2011	Density of population per acres	% of Literacy	Percentage of total working force of Cultivator	Agricultural Labours	Other worker	% of net sown area	Land per share	% of uncultivated area	% of unirrigated areas	% of irrigated areas
1. Bangaro (Tundi)	541.93	694	1.28	59.59	18.12	72.48	2.73	24.19	0.78	11.91	90	10
2. Behachia (Govindpur)	516.94	788	1.52	55.83	31.86	5.7	16.71	47.48	0.65	9.80	96	4
3. Tetulia (Nirsa)	541.93	1666	3.07	59.36	12.27	26.62	19.82	24.19	0.32	11.51	90	10
4. Paharpur (Baliapur)	529.77	1819	3.43	64.04	11.29	18.21	21.75	34.84	0.29	26.30	95	5
5. Saldaha (Baghmara	237.07	334	1.40	63.87	11.11	37.03	37.77	52.55	0.70	9.47	96	4
6. Kharni (Topchanchi)	456.98	951	2.08	48.89	6.63	20.68	16.13	54.67	0.48	10.26	96	4
7. Salgudari (Dhanbad	253.38	5450	21.45	68.62	0.59	3.99	13.95	19.75	0.04	59.22	98	2
8. Rohraband (Jharia)	1152.91	8248	32.60	89.47	0.90	0.20	92.51	11.65	0.13	65.01	99	1

5
Chapter

Analysis of Change in Land Use Pattern

ANALYSIS OF CHANGE IN LAND USE PATTERN

Land is a basic natural resource. It plays a significant role in determining man's economic, social and cultural progress. In this chapter an endeavour is made to examine the decadal variation of land of Dhanbad district.

I. VOLUME OF CHANGE IN FOREST LAND USE :

In Dhanbad land use of forest decreased during 4 decades. 1960-61 about 15% of total geographical area was covered by forest land while 2014-15 year it was reduced to 9.27%. In terms of areas in 1960-61, 106407 acres land was under forest. But now it has decreased to 46773 acres.

The greatest deforestation was during 1960-61 to 1980-81. In 1960-61 forest cover was 14.99% of the total land. It decrease to 8.38% in 1970-71 and in 1980-81 it was just 7.64%. The reason behind this decrease is ruthless exploitation of forests. This was because of increasing population pressure in the region. With the increase in population the demand for timber increased and deforestation was the result. The deforestation should be practiced but only matured trees should be cut and it should be practiced along with afforestation.

But study of forest land during 1980-81 to 2014-15 reveals the fact that it is increased by 1.5%. This increase has been achieved due to the efforts of the state government with launching the scheme of afforestation and plantation of new crop Appreciating the necessity of developing forest, the government of India declared its forest policy in 1952. According to this policy, it was decided to raise steadily the

Table - 5.1
Land Use Pattern (1960-61 to 2014-15)

Sl. No.	Land Use Pattern	1960-61		1970-71		1980-81		2014-15	
		Area (acres)	Percentage	Area (acres)	Percentage	Area (acres)	Percentage	Area (acres)	Percentage
1.	Forest	106407	14.99	-	8.38	-	7.64	-	9.27
2.	Land Pur to Non-Agricultural Uses	178917	25.20	-	24.711	-	24.05	-	23.68
3.	Barren & Uncultivated Land	70310	9.90	-	11.84	-	13.67	-	15.96
4.	Permanent Pasture and Other Grazing Land	6346	0.89	-	1.02	-	0.82	-	0.28
5.	Land Under Miscellaneous Tree, Crop and Groves	6320	0.87	-	0.89	-	0.88	-	1.05
6.	Culturable Waste	24893	3.46	-	7.85	-	6.25	-	5.58
7.	Current Fallow	66503	9.36	-	7.92	-	11.95	-	12.23
8.	Other Fallow	92930	8.86	-	8.78	-	12.63	-	1.54
9.	Net Area Sown	193764	27.29	-	28.61	-	22.11	-	20.62

Source : Calculated by the Author

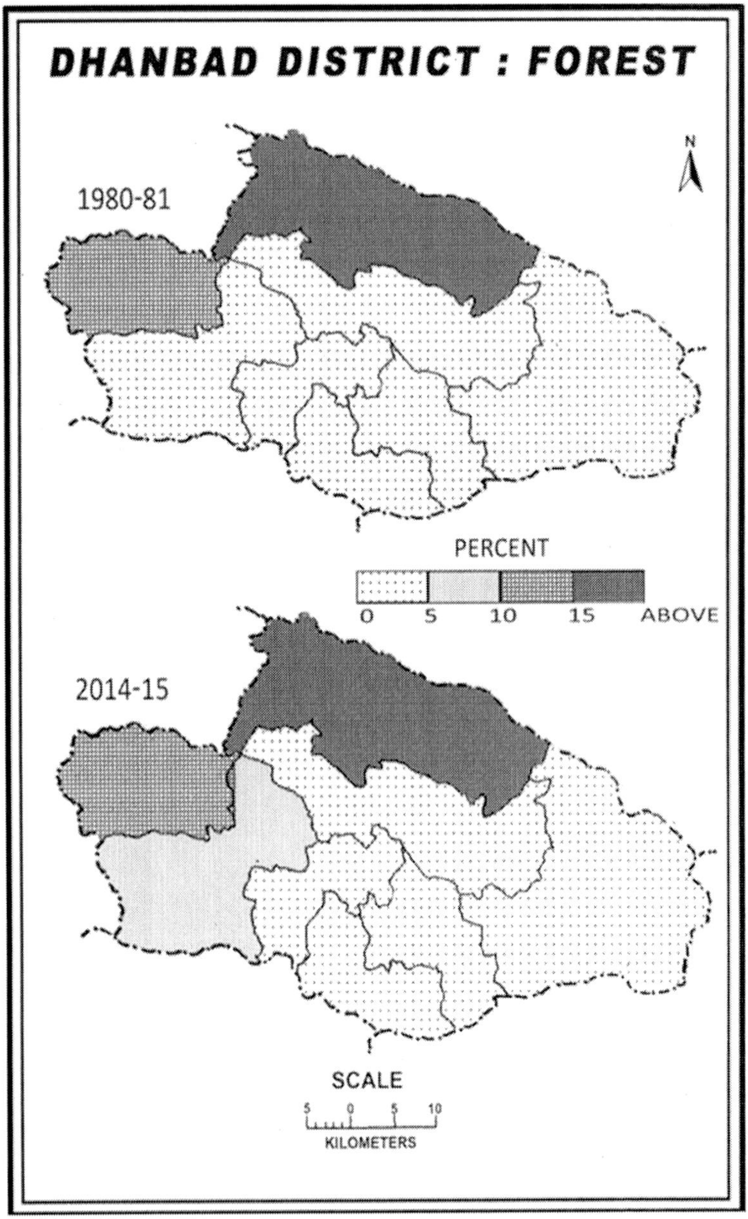

Fig. 5.1

Source : The Map Prepared by Author

area under forest to 33% for the century as a whole.

An increase in forest land is noticed in 5 blocks in varying proportions. The highest increase was registered in Baghmara (3.58%), Dhanbad (1.86%), and Topchanchi (1.85%). Rate of increase in Baliapur and Nirsa block is less, while in Govindpur and Tundi block forest land has decreased. Jharia Anchal is forest less where all forest land is converted into other use of land due to development of mining and industries.

II. VOLUME OF CHANGE IN LAND PUT TO NON AGRICULTURAL USE :

The amount of this category of land has not increased during last 40 years due to a huge demand for land for settlement, industries, rail and road network. Data shows that where in 1960-61 non agricultural use available was 25.20, it came down to 23.68 and in 2014-15 of the geographical area. In 1971 it was 24.05%. Main cause for these decreases may be that other kind of land is barren land and was used for settlement and industrial proposes.

But during last 20 years distribution of non agricultural land is even. In 1980-81 non agricultural land was occupied in Jharia (75.46%), Dhanbad (31.38%), Govindpur (31.38%), Tundi (13.09%), and Topchanchi (16.62%) while it is increased to 84.24%, 32.26%, 21.50%, 10.45% and 22.27% respectively in 2014-15. Largest increase was shown in Jharia Anchal (8.78%) due to mining activity and increase in population. Most part of Jharia falls in non agriculture land. Several coke plants are established in Govindpur Anchal. It attracts number of people so the demand in non-agricultural land has increased. Due to urbanization in Dhanbad, Topchanchi Anchal the amount of non agricultural land has increased. Data shows a decrease in Baliapur, Nirsa and Baghmara for non-agricultural land because of transfer of land from one category to another.

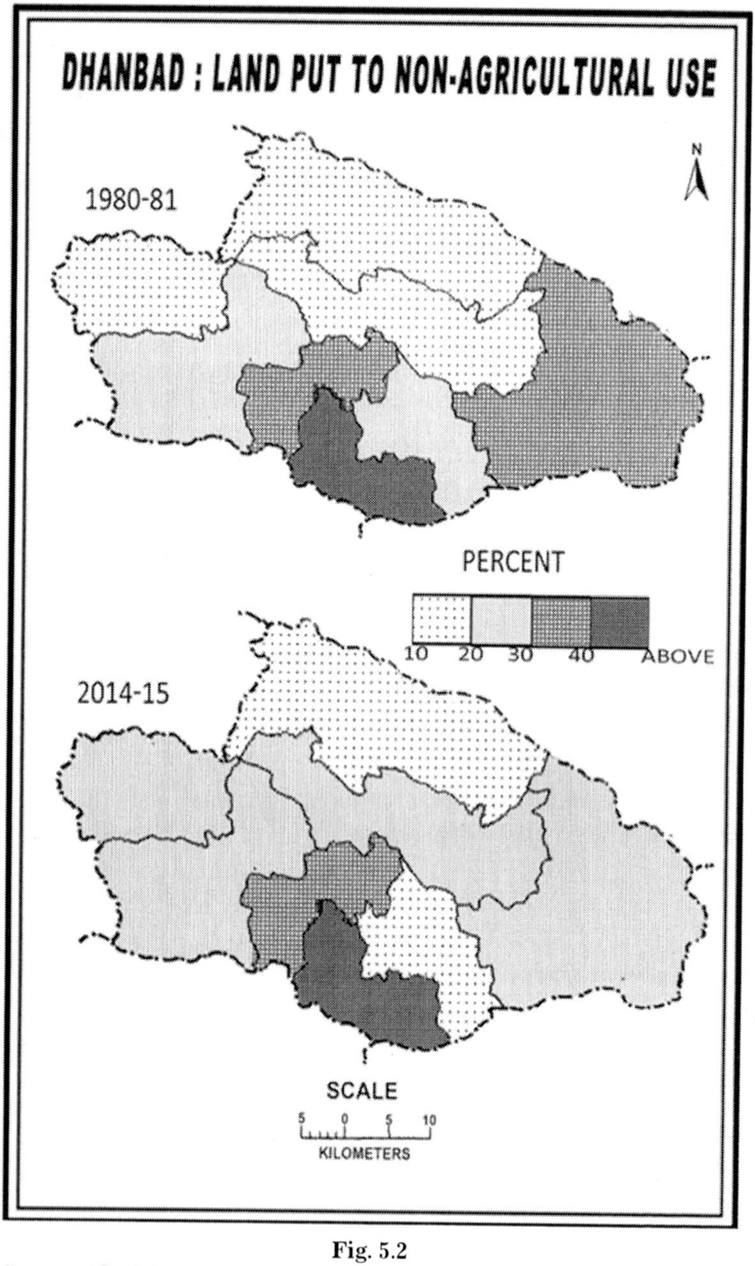

Fig. 5.2

Source : The Map Prepared by Author

III. VOLUME OF CHANGE IN BARREN & UNCULTIVATED LAND :

Between 1960-61 to 2014-15 percentage of barren and uncultivated land has increased from 9.90% to 15.96%. This category of land covered 11.84% in 1970-71 as 13.67% in 1980-81 of total geographical area. It means during 4 decades barren and uncultivated land has a major share in the total area. It is a result of deforestation, gully erosion and old mine land.

Increase in this type of land is not even in all blocks. In percentage the increase is 3.5%, 2.67%, 0.24% and 0.56% respectively. While in comparison from 1980-81 to 2014-15 barren land and uncultivated land decreased in Tundi, Baliapur Anchal.

Increase in barren land is a matter of anxiety. Today the district is suffering from food scarcity so it is necessary to stop extension of this type of category of land. Barren land can be utilized by planting trees (it helps to control pollution, helps in balancing the ecology). There is a large scope of industrialization in this district so the barren land should be utilized to establish new industries. Planning is necessary to stop the increase in barren land.

IV. VOLUME OF CHANGE IN CULTIVABLE WASTE LAND :

This type of land also increased during 1960?61 to 2014-15. In 1960-61 only 3.46% land of the total land was under cultivable waste land. It increased to 5.58% in 2014-15. There was a large increase in the 70's of 7.85% land covered as cultivable waste. In term of acres 1960-61, only 24893 acres land was barren and cultivable waste increased to 57958 acres. Difference shows that it practised more than double in 10 years. But increase in the population of land under this use has decreased to 6.25%, in 1980-81 and 5.58% in 2014-15. Main cause of this improvement is in irrigation facilities. Govt. Of Jharkhand is distributing load to the people for well and tank construction to achieve more irrigated land. In this

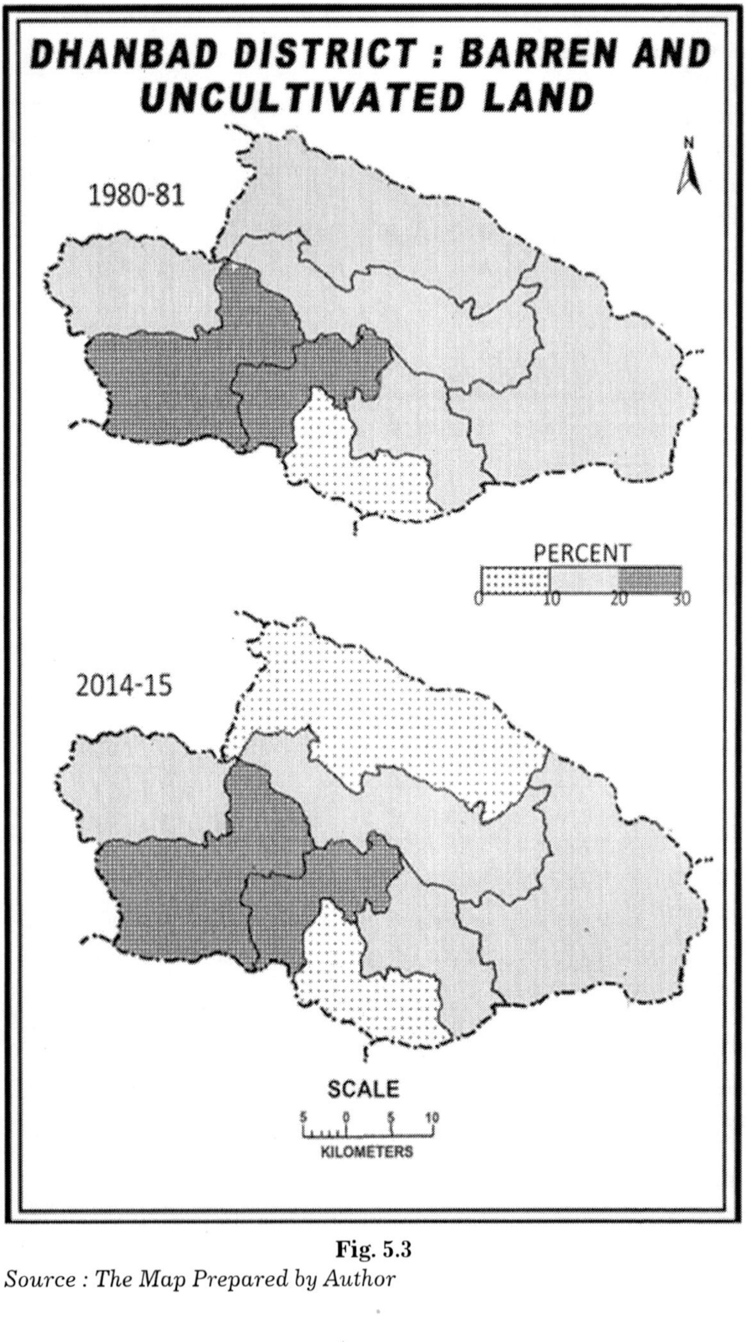

Fig. 5.3

Source : The Map Prepared by Author

concentration all small river should also utilise for irrigation to make check dam on it. Distributions of barren and uncultivable waste land in all bocks are uneven and shows change in past years. There were five Anchal viz. Tundi, Nirsa, Baghmara, Topchanchi and Baliapur where percentage of this type of land was 7.15, 7.65, 7.70 and 6.42 percentages respectively in 1980-81. It was more than regional percentage. During that period Dhanbad, Jharia had less land under this type due to development of trade & commerce. A comparison between 1980-81 to 2014-15 shows that this type of land has increased in Jharia by 1.92% and Topchanchi by 0.37% and Dhanbad by 0.84 percent.

V. VOLUME OF CHANGE IN FALLOW LAND :

Between 1960-61 and 2014-15 the percentage of fallow land has increase by 5.33%. In terms of areal extent the increase is of 11810.6 acres. The magnitude of increase in the fallow land is not ideal index for cultivation.

Land under this category in 1980-81 was as follows. Govindpur (30.95%), Dhanbad (24.56%), Baliapur (24.91%), Topchanchi (30.49%), Nirsa (23.55%), Baghmara (21.31%). This data has changed in 2014-15. Now Nirsa holds largest share of its total area 33.29%. Other blocks viz. Dhanbad, Baliapur, Govindpur, Tundi, Baghmara and Topchanchi covers 20.65%, 26.56%, 26.48%, 18.70% and 21.6% respectively under fallow land. While Jharia block holding the lowest share of mere 3.09% under this type. The reason behind this is that least cultivated area. Amount of fallow land is varying due to uncertainty in rainfall or monsoon. The whole region is depended upon Rain God for cultivation. Improvement in irrigation facilities, fallow land may utilize for crops.

VI. VOLUME OF CHANGE IN NET SOWN AREA :

The variation in the concentration of net sown area is the district showed decrease about 7% during the period from 1960-61 to 2014-15. 1960-61 total sown area was 27.29% of the total geographical area of the district. It has reduced to 20.62% in 2014-15. In terms of acres the sown area in 1961 was 193764

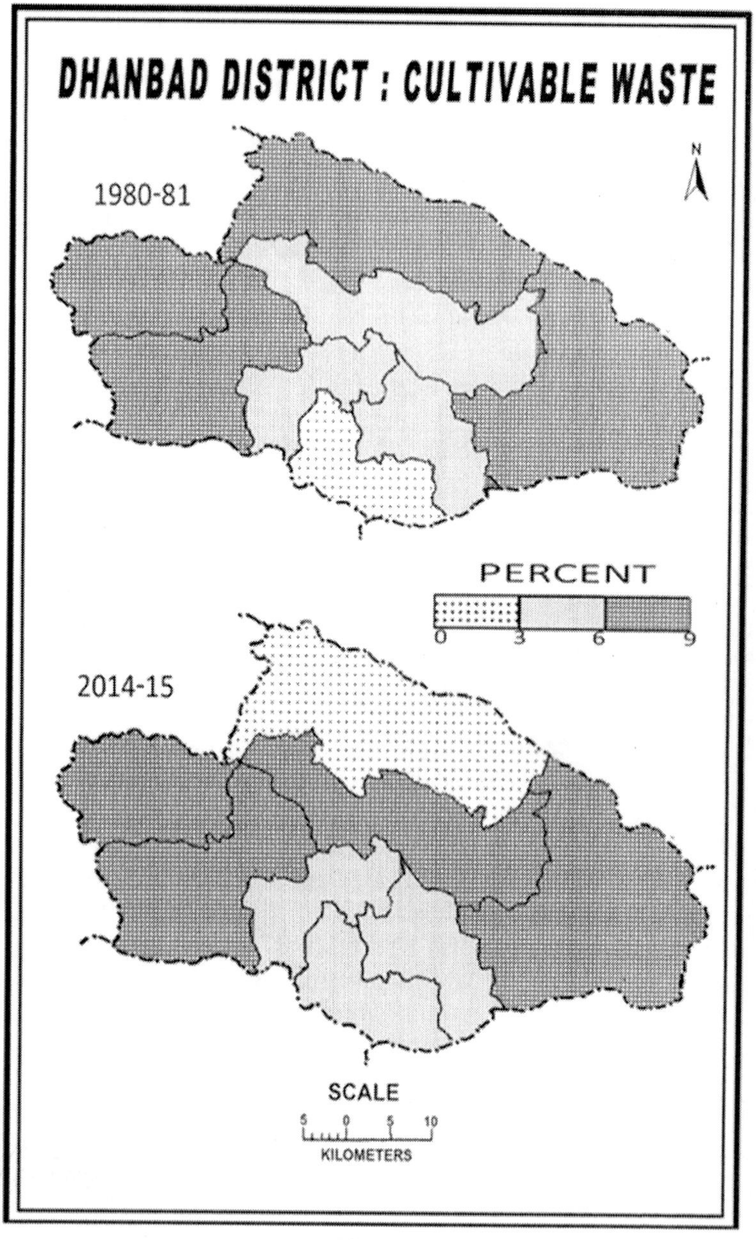

Fig. 5.4

Source : The Map Prepared by Author

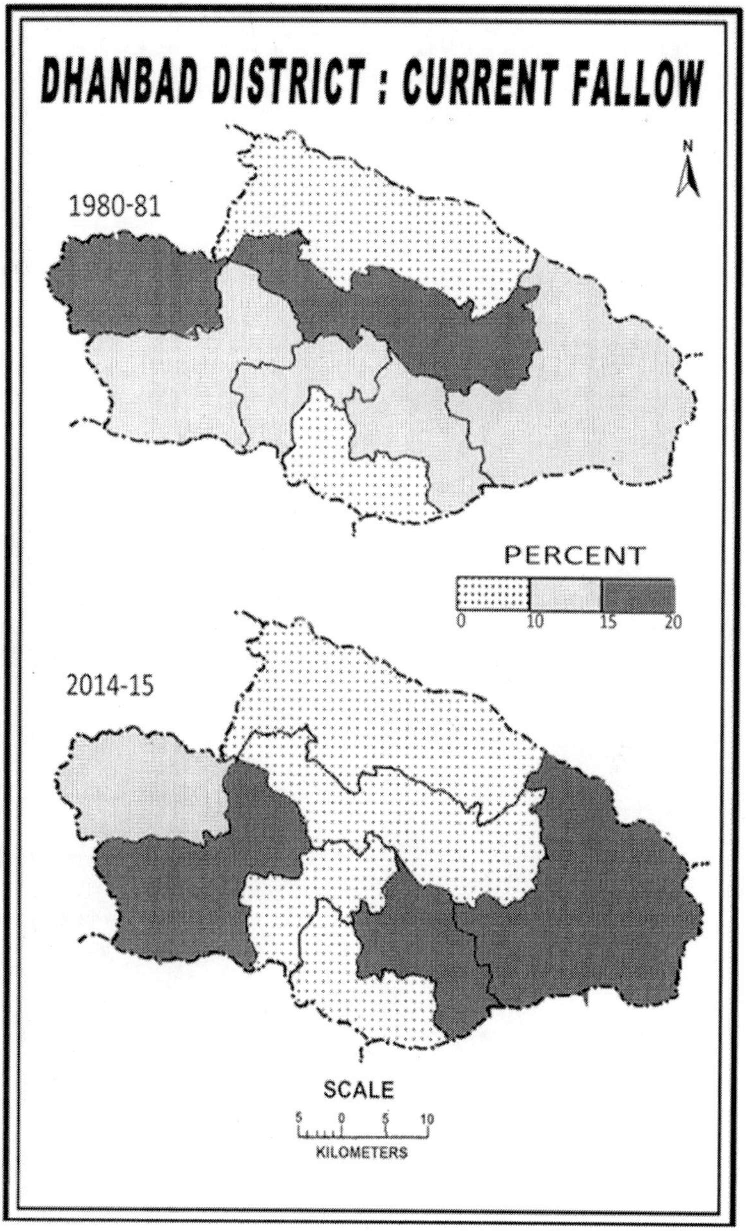

Fig. 5.5

Source : The Map Prepared by Author

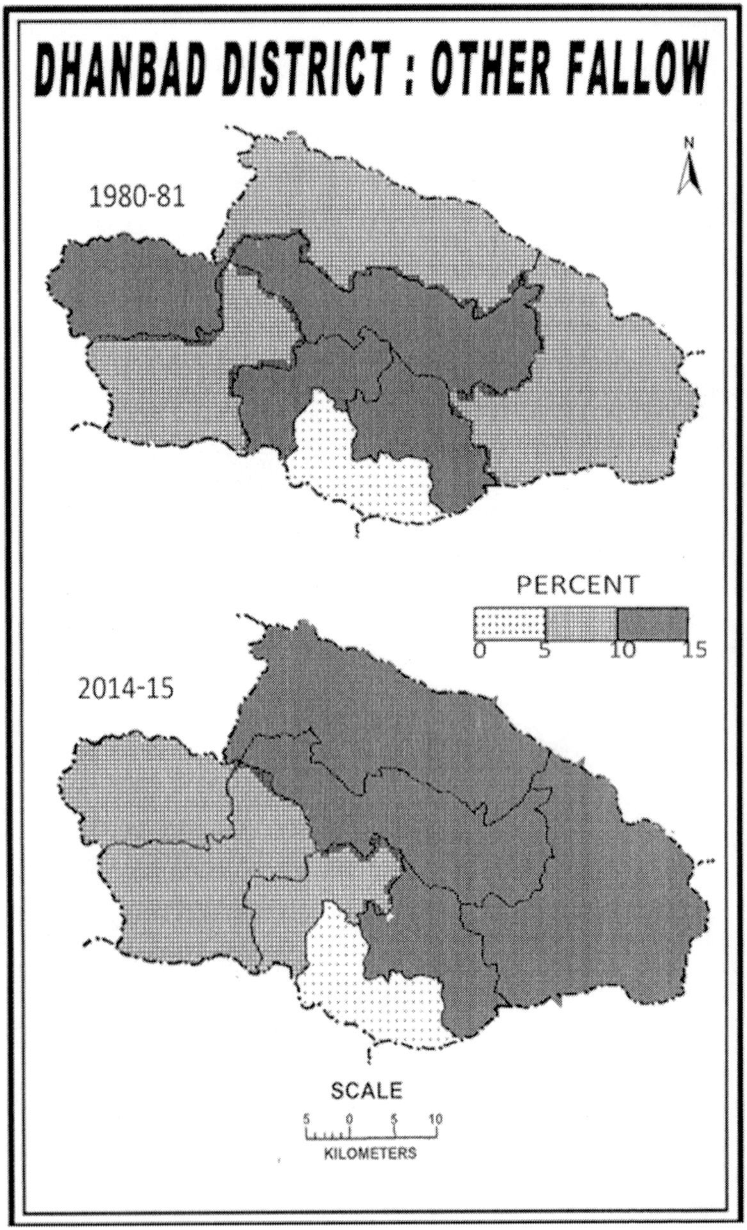

Fig. 5.6

Source : The Map Prepared by Author

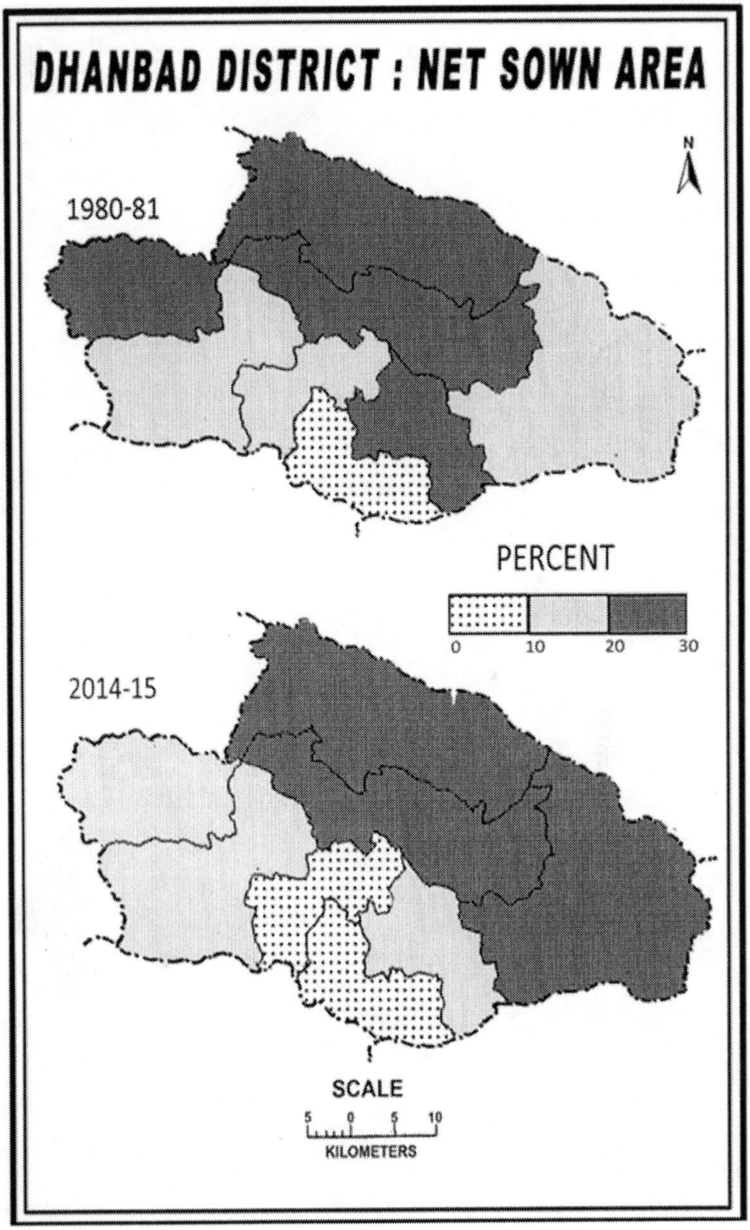

Fig. 5.7

Source : The Map Prepared by Author

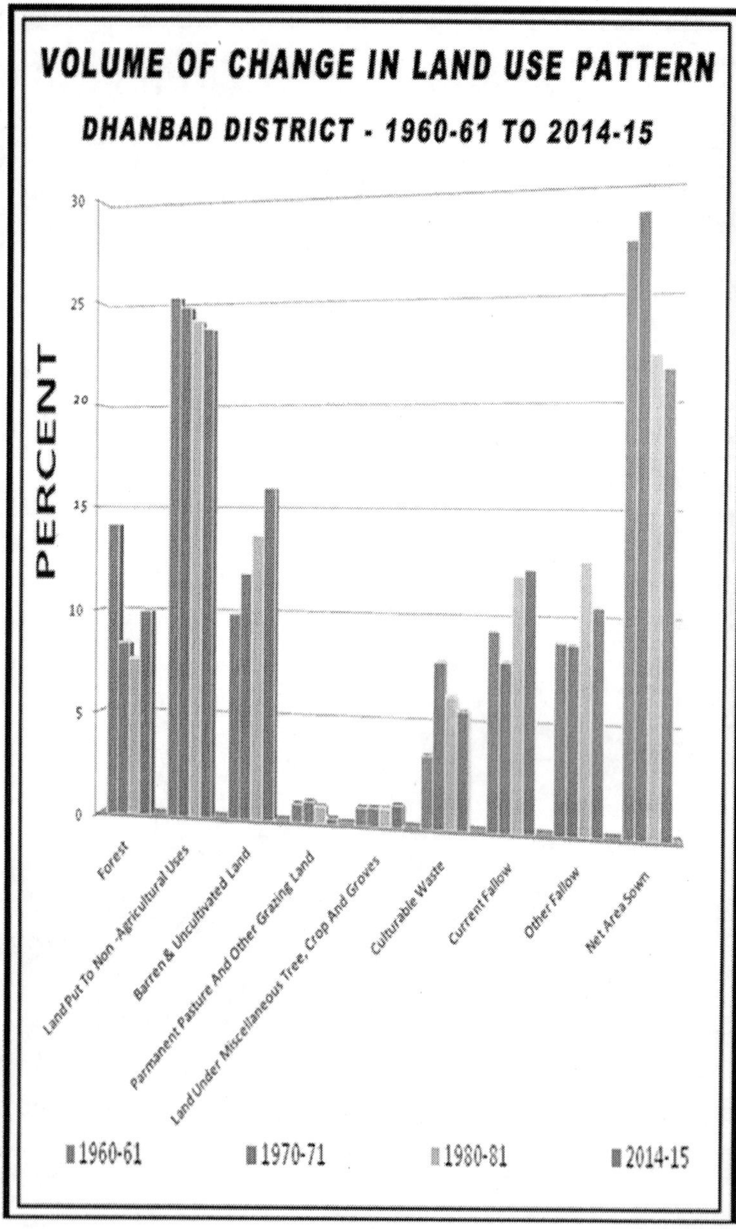

Fig. 5.8

Source : The Map Prepared by Author

acres while in 2014-15 it decreased to 104052 acres. It was 28.61% of total geographical area in 1980-81. The ratio of net sown area to the total area of the district varies considerably from block to block. In 1980 Govindpur, Baliapur, Tundi, Topchanchi and Baghmara, Nirsa and Dhanbad has 31.44, 26.86, 26.60, 17.02, 16.19, 11.27 land under net sown area. It shows that Topchanchi, Govindpur, Tundi and Baliapur block had more area than the regional average. Other blocks are containing less than regional average decreasing rate of sown area also varied in 2014-15 in all blocks. During this year Tundi has more land under this category and Jharia Anchal has mere 1.64% of its total area. In other blocks the figure is as follows- Baliapur 27.50%, Govindpur 26.45%, Nirsa 13.87%, Baghmara 18.93%, Topchanchi 2.35% and Dhanbad 5.54%. 1997-98 four block viz. Tundi, Govindpur, Topchanchi and Baliapur have more than ratio of district and Baliapur have more than ratio of district and other from has less sown area than the district level. Less amount of precipitation increasing Culturable waste and fallow land.

6
Chapter Summary and Conclusions

The district of Dhanbad is situated between 23°37'3' to 24°40'N latitude and 86°6'30" E to 86°50'E longitudes in the eastern part of the Jharkhand. River Barakar constitutes the northern boundary of the district, river Damodar constituted the southern boundary, Jamunia river marks western boundary of Dhanbad. The district is bounded on the west by Bokaro, on the north by Jamtara and Giridih district, on the east by Bardwan district, which is a portion of West Bengal, and on south by Bokaro. It is about 75 Kms Long from east to west and 45 Kms Wide from north to south. It covers an area of 504094.97 acres or 2040 sq.km and has a population of 2664487 (2011).

From a geological point of view Dhanbad can be divided in to two parts, northern half which consist entirely of ancient-crystalline rocks and the southern occupied by the great coal basin of Jharia in the west and the cis- Barakar portion of the Raniganj coal basin in the east, with intervening area of crystalline rocks. The Jharia coal fields cover basin in the east, with intervening area of rocks. The Jharia coal fields cover 175 sq. Miles of which 105 are occupied by coal bearing rocks. The oldest geological formation of Dhanbad district belongs to Dharwar system. In these formations are found rock types of both sedimentary and igneous origin. The recent formations which form superficial deposits are the alluvial sand of river and stream. Thick deposits of recent alluvial sand occur adjacent to the bank of Damodar and Barakar River and the Jamunia, Khudia and Pusai stream.

Dhanbad district forms the eastern part of the Chhotanagpur pleatue. But for some portions in the west, it has an average elevation of 500-1000 ft. above the mean sea level. Only the Dhangi Hills rise to over 1000 ft. Towards the east, the valleys of the Barakar and the Damodar are less than 500 ft. above mean sea level. There are three broad natural divisions in the district (a) The Hilly region in the North West. (b) The upland contains coal mines and (c) The remaining upland and plains.

In the western middle part of the district are the Dhangi hills, situated in the strip falling between the Grand Trunk Road in the north and the Grand chord line of the eastern railway in the south. These hills extend from Pradhan Khanta to Govindpur, reaching a maximum altitude of 1256 ft. at Dhangi. Further north a branch of Parasnath hill runs through Topchanchi and Tundi the highest point of 1500 ft. being reached at Lalki. The southern part of the district is largely comprised of undulating land. The general slope is from west to east, the direction followed by the two major rivers Damodar and Barakar.

The Damodar is the most important river of Chhotanagpur plateau. It enters Dhanbad district at its confluence with the Jamunia, a stream which marks the western boundary of Dhanbad. Further east, the Damodar is joined by the Katri, which rises in the foot hills below Parasnath and traverses through the coal field area. The Damodar flows for 48 miles through the district being joined by Barakar at its eastern border near Chirkunda. The Barakar, which forms the northern boundary, traverses a distance of about 48 miles in the district. It flows in south westerly direction up to Durgapur and then south till it joins the Damodar near Chirkunda. The Maithon dam is located on this river. Among other small rivers in the district are the Pusai, the Ijri, the Khudia and the Katri. All rivers of the district are seasonal in character and remain dry during the summer season.

The climate of the district is characterized by general dryness. It is very pleasant during the cold weather from Nov. to Feb. There after the climate becomes warm. It remains hot until the monsoon breaks towards the middle of June with the setting in all remains the temperature falls and humidity rise. July to Oct. are the rainy months. July and August are the wettest months. The average rainfall during July is 341 mm. and that in Aug. 313 mm. The average annual rain fall in the district is approximately 1300 mm. There is hardly any rain during Dec. to Feb. and the average for the three month from March to May is well below 92 mm. The region suffers much from vagaries for monsoon. Scarcity of rainfall cause problem for proper agricultural land use and may results drought. Climates determine the activities of farmers and control the production of crops to a great extent. The farm operations are also governed by the favourable weather conditions.

The district has rather poor and infertile soil largely comprising of laterite, alluvial deposits occur at the confluence of the Damodar and Barakar River. The soils of the district are broadly classified into four groups according to texture (a) stoney and gravely: - it is a low grade soil near the foothills. (b) Sandy soil: Locally known as Bali, found in the river. (c) Loamy soil- it is found near the lowland and formed by rain washing from higher area. Agriculturally these soils are best adapted for cultivation. They are suited to every kind of crops. This soil is put under paddy mainly. (d) Clayey soil found near the tanks and the river. It is very productive.

Paddy is the main crop. The soils for rice cultivation of the district have been differentiated by the cultivator on the basis of their position into three classes' viz. Baad, Kanali and Bahal. Baad lands are upland made by high terraces and depending entirely on the rainfall.

Kanali lands are situated between Baad and Bahal land. The soil is loamy.

For obtaining a good yield organic manures and fertilizers should be added. Bahal lands are the best type of land for growing paddy. They are low lying and they receive enough water through percolation and it is retained. The soil is fertile and rich in humus. For better production, use of fertilizer and manures are added. Sometime high lands situated over

Baad land are called by cultivators Tanr land. They are fit for Ragi, Urad and Oil seed. These lands are subjected to soil erosion. Bounding properly and checks for soil erosion are necessary for these lands. Land adjoining to homesteads is known as Bari land. These lands are best suited to vegetable.

According to 2011 census there are 26 lakh persons over an area of 2040 sq km having a density of 1316 people per square km.

Population of the district growing fast rate. In the decade 1961-2011 population experienced an increase of 266.83%. The growth of population is not uniform throughout the region. Anchal Jharia, Dhanbad and Nirsa experienced much more than other Anchal. In these Anchal the increase is chiefly due to influx of immigrants from other part of Bihar and Uttar Pradesh. Per capita land is decreasing day by day at a great pace. The existing industries are absorbing a large portion of population of the working population and hence the burden on agriculture is much less than in other district of Jharkhand.

So far as density of population is concerned this district can be divided into two zones Industrial zone and Agricultural zone. The industrial zone contains 20% of total of the total geographical area of the district, and accounts for more than 45% of total population. The agricultural zone which is 80% of the area account for 55% of population. According to 2011 census majority of people are engaged in non agricultural activities (77%) and more than 23% of the total worker are engaged in primary sector. The region has 20.37% people engaged as main worker and out of main worker cultivations and agricultural labours are 11.50% and 8.28% respectively. The region has only 3.07% person engaged in household industry.

Taking Dhanbad district as a whole consideration 45.22% of its total area can be said to be arable while 54.78% is not available for cultivation purpose. But out of the total percentage of available land only 20.62% was actually sown with crops and others are under fallow land. High percentage of fallow land is result of lack of irrigation facility. Agriculture depends upon the mercy of Rain God. So far as agriculture is concerned.

Dhanbad is mainly a rice producing tract. About 95% of cropped area is under paddy cultivation. A few percent of land is put to under production of wheat, sugarcane, maize, gram and vegetables.

CONCLUSION

The detailed analysis of geographical and socio-economic conditions influencing land utilization, crop production and associated problem of Dhanbad district has enabled the investigator to draw following finding conclusions

1. Dhanbad district being in the eastern portion of Chhotanagpur pleatue is an industrial region where majority people are engaged in non-agricultural activity.
2. Cropped land has decreased during 4 decade. Now only 20.62% of total geographical area is under net sown area.
3. The region has made considerable progress in literacy as well as level as education. In fact most of the village has considerable number of graduates. Even female education has remarkable accelerated.
4. Occupational structure has also considerably changed due to the impact of the industrialization urbanization, education and modernization. Now people are diverting toward secondary and tertiary occupation.
5. Impact of scientific innovation is obvious in the field of varieties of paddy production. Although farmers have not adopted scientific method of cultivation, Morden agriculture equipments. But they have paid much attention towards new verities of seed (e.g. HYV High yielding Verities seed).
6. Due to lack of irrigational facilities land under area sown more than once is only 1%.
7. Almost all villages have experienced considerable growth

of population due to decrease in death rate and also decrease in infant mortality. Birth rate has not decreased to a considerable extent. Medical facilities have enabled the average age of the people.

8. Rate of increase in population has been faster in urban area. This is due to the development of mining and manufacturing industry in some part of the district. Urbanisation in this district has been essentially a town ward migration especially of the male population.

9. Growing insecurity due to naxal activities has compelled several rich farmers to shift the attention from agriculture to non agricutural activities. Most of these farmers started disposing of their land and investing their capital in urban center.

10. The surrounding area of urban center has experienced remarkable growth in the production of vegetable.

11. In the study area forest land is in decreasing tendency. In 1960-61 about 15% of total geographical area was covered by forest now only 9% occupied by forest.

12. The study area is a plateaus and having uneven land surface. All river and streams are non-perennial they do not contain water for irrigation other than in rainy season.

13. Dhanbad is the most highly industrialized district in Jharkhand. The dominant industry of the district is coal mining which has attracted and brought about a concentration of numerous other industries.

14. The district is well served by railways and roads. The district is easily connected with Kolkata, Jamshedpur, Ranchi, Patna, Delhi, Mumbai and other cities.

15. In the study area coal is the most important article of export. Export goes to Banaras, Delhi, Kolkata and other cities.

16. Study area has to import food grains because of low rate of productivity and inadequate ratio of net sown area.

Bibliography

BOOKS

Agrawal, N.C. (1961), The Food Problem of India, Bombay.

Ahmad, E. (1965), Bihar : A physical economical and regional Geography, Ranchi University, Ranchi

Ahmad, E. (1976), Some aspects of Indian geography, central book depot.

Allen, R.G. (1940), Outline of Indian agriculture, Baroda.

Amani, K.Z. (1976), Agricultural land use in Aligarh District, Kumar Publication, Aligarh

Anakeri, M.R. Etal., Soil management in India.

Baker, O.E. (1973), A graphic summary of physical features and land utilization in U.S.A., U.S. department of agriculture: miscellaneous publication.

Bansil, P. (1977), Agriculture Problems of India, New Delhi.

Banerjee, C.P. (1969), Essays on Agricultural Geography, Nahaskti Press, Calcutta.

Barnes, C.P. (1936), Land classification objectives and requirement, U.S. resettlement administrative land utilization division, land use planning publication.

Bartholomew, H. (1955), Land use in American cities, hardvard university press, Cambridge, mass.

Balkrishnan, R. (1959), Techno-economical survey of Bihar, Asia publishing house, New Delhi.

Beaver, S. H. (1943), Soke of Peterborough: the report of the land utilization survey of Britain, (edited by L.D.

stamp), London.

Best, R.H. and The changing use of Land in great Coppock, J. T. (1962), Britain, Faber and Faber, London.

Bhat, L.S. (1972), Regional Planning in India, Statistical publishing society, Calcutta.

Bhatia, B.M. (1970), India's food problem and policy since independence, Semaiya, Bombay.

Buck, J. L. (1937), Land Utilization in China, University of Chicago press, chapter VI.

Buck, J. L. (1956), Land Utilization in China, the council and Central Affairs, New York.

Burns, W. (1941), Sons of the soil: studies in Indian Cultivators, Delhi.

Chauhan, D.A. (1966), Studies in Utilization of Agricultural Land, Educational Publishers, Agra.

Chen, C. S. (1950), Land Utilization in Taiwan, National Taiwan University, Taipei.

Chisholm, Michael (1962), Rural Settlement and Land Use, London.

Chouridule, P.B. (1971), The "Land use in chichgarh", 58 session of Indian Science Congress, part III.

Chatterjee, S.P. (1968), Progress of Geography in India during the past fifty years. Indian Science Congress Association, Calcutta.

Chatterjee, S.P. (1969), Land Use Characteristic Around Bodh Gaya, Indian Science Congress Association fifty-sixth session, Bombay (Powai), part III, section of Geology and Geography.

Chatterjee, S.P. (1948), Land Utilization in the District of 24 Parganas, Bengal, Calcutta.

Chatterjee, S.P. (1956), "Preliminary remarks on Land Use Survey in India", Proceeding, International Geographical

Seminar, Aligarh.

Chatterjee, S.P. (1958), Land Utilization Survey of Howrah District

Chatterjee, S.P. (1962), Land Use Survey in India Proceedings, Summer School in Geography, Shimla.

Claart. C. (1967), Population Growth and Land Use, Macmillan and co. London. Coppock jt and best rh 1962 the change use of land in Great Britain Faber and Faber London.

Coppock, J.T. (1970), An Agricultural Geography of Great Britain, London.

Dayal, P. (1953), Bihar in Maps, Kusum Prakashan Patna.

Davis, K. (1950), The Population of India and Pakistan, Princeton.

Davidm N. L. (1961), Land Use and Opulation in Tobago, London.

Demetris, C. (1959), The Evolution of Rural land Use Patterns in Cyprus England.

Diwakar, R.R. (1959), Bihar through the ages orient Longmans Calcutta.

Faulkner, C. N. (1951), Pemberton Valley Land Utilization Survey British Columbia Department of Land and forests Vectoria.

Finch, V.C. and Geography of the worlds Agriculture

Baker O.E. (1917), Washington.

Forde, C. D. (1966), Habit, Economy and Society, Methuen and co. London.

Freeman, T.W. (1967), Geography and Planning Hutchinson and co. Ltd. London.

Ganguli, B. N. (1938), Trends of Agriculture and population in Ganga Valley London.

George Kay (1965), Changing Pattern of Settlements and land use in the Eastern Province of Northern Rhodesia University of Hull Publication Yorkshire.

Graham, E.H. (1944), Natural Principles of Land Use, Oxford University Press, London.

Gray, L.C. (1923), The Utilization of our land for crops pasture and forest year book of us department of Agriculture, New York.

Gyanchand (1956), Some aspects of the population problem of India, Patna.

Hill ga ed (1960), A multiple land use plan for the Glackmeyer Development area Ontario.

Hundson, G.D. (1935), The Rural Land Classification Programme PVA Washington.

Jackson, J. N. (1966), Surveys for town and country planning, London

Jain, S. C. (1966), Changing Indian Agriculture Vora and Co, Bombay

Josay, Y. G. (1972), Agricultural Geography of the Marwadda Basin MP Hindi Granth Academy

Kish Leslie (1965), Survey Sampling John Wiley and sons INC, USA.

Kostrowicki, J. (1962), Land Utilization : Method and problems of research, Warsazawa.

Kostrowieki, J. (1965), ED Land Utilization in East Central Europe : Case Studies, Warsazawa.

Krishnan, M. S. (1956), Geology of India and Burma Higginbotham, Madras

Kumar lsss et al (1962), Agriculture in India vols. I, II and III Asia publishing house, Bombay Land (1958), The year book of Agriculture 85 Congress no. 280 Washinton pp. 363-64

Learmonth ATA (1962), Sample Village in mysore state, India Geographical study, Liverpool

Leboe, jhg (1955), Land Use surveys in the Sudan Natural Resource food and population of inter-tropical Africa report of symposium September p. 77.

Leslie, S. (1960), Land Use in Northern Ireland, London.

Madan, G.R. (1959), Changing pattern of India villages, S. Chand and Co., Delhi

Mandal, R.B. and Recent trade and concept in Geography

Sinha, V.N.P. (1980), Vol. II Concept publishing co. New Delhi

Mandal, R.B. (1979), Reading in land reforms research publisher Monghyr.

Mellor, J. W. (1969), The Economic of Agricultural development Cornwell University press, New York

Misra, R.P. (1968), Diffusion of Agriculture innovations, Mysore

Mohammad, Ali (1978), Studies in Agriculture Geography Rajesh Publications, New Delhi

Mohammad, Noor (1978), Agricultural Land Use in India, Inter India Publication, New Delhi.

Monkhouse, F.J. and Wilinsoon, H.R. (1952), Maps and Diagrams London. Moreland wh 1912 the Agriculture of UP Allahabad.

Food planning for four hundred Millions Macmilan and Co., London.

Applied Geography in Hungary Budapest. Implementation of Land Reform in Bihar since Independence

Farmer of India, Vol.-I, 1961 ICAR, New Delhi.

Soil of Bihar ICAR New Delhi Ram Das SK weather and Indian farmer

The Agricultural problems of India the Army Press, Lahore

Soil Conservation and India Indian Council of Social Science and research, New Delhi.

The Sahi Valley irrigation Projects Chicago Land Utilization of Eastern UP Aligarh

A Survey Research in Geography NCERT Publication New Delhi.

Agricultural Geography Tara Publication, Varanasi.

Singh, Jasbir (1974), Agricultural Atlas of India: A Geographical Analysis Vishal Publication Kurukshetra.

Singh, R.L. (1967), Practical Geography, Banaras

Singh, R.P. and Kumar, A. (1970), Monograph of Bihar Bharti Bhawan, Patna.

Singh, S.P. (1971), Dholari Village : Rural Studies in Geography series T Silver Jubilee Publication Varanasi

Singh, V.R. (1970), Land Use Pattern in Mirzapur and Environs Banaras

Singh, Jagdish(1964), Transport Geography of South Bihar BHU Press Varanasi

Stamp, L. D. (1950), The Land of Britain its use and misuse Longmans green and Co. Ltd. London.

Stamp, L. D. (1951), Land for tomorrow the under developed world Bloomington

Stamp, L. D. (1960), Our Developed World Faber and Faber London

Stamp, L. D. (1960), Applied Geography Penguin book INC USA.

Stapledon, R.G.S. (1962), The Land : New and tomorrow Symons 1967 Agricultural Geography g bell and sons ltd. London. Ueno f 1960 land utilization in Japan Tokyo

Vink, A.P.A. (1975), Land use in advancing Agriculture Springer Verlog, New York.

Wandia, D.N. (1961), Geology of India Macmillan and Co., London.

Wandia, D.N. and Geology and Structure of Northern India

Auden, J. B. (1939), Memories of the Geological survey of India, Vol.-73, Delhi.

Land Utilization in Australia Victoria Warriner d 1967 Land Reforms in principles and practices Crandon Press Oxford Wynne thorne 1963 land and water use Washington

Crop Land Use Pattern in Rajasthan Conference of Geographer, Bhilwara.

JOURNALS

Ahmad, E. (1954), The Soil of Bihar Geographer, Vol. VI, pp 7-32

Ahmad, E. (1959), Bhusur Village : A Sample survey of Land Use Pattern Bombay Geographical Magazine, pp. 45-62

Ahmad, E. (1959), Scope of Land Utilization Survey in India, Indian Geographical Journal, Madras no. 1

Banerjee, B. (1969), Infra Structure of Indian Agriculture, Essays in Agricultural Geography, p. 260

Banerjee, B. (1964), Changing Crop Land of West Bengal, Geographical review of India vol. XXIV

Chaterjee, S. P. (1941), The Place of Geography in National Planning, Calcutta Geographical Review, vol. 3

1. Afforestation in a barren land, village Paharpur

2. Paddy Crop Plantation in Monsoon Season

3. Cattle is Grazing in the field

4. Ploughing of the Agricultural Field

5. Harvesting of the Paddy Crops

6. Research Scholar with the Supervisor in a sample village

7. Jharia Coal Fire Area in Dhanbad district

8. Open cast Mining in the Jharia Anchal

9. Maithon Dam - A Tourist Place

10. Natural Beauty of Bhatinda Fall - Dhanbad